The Owl Handbook

"Owls have been both feared and venerated,
despised and admired, considered wise and foolish,
and associated with witchcraft and medicine,
the weather, and births and deaths—and have
even found their way into haute cuisine."

—HEIMO MIKKOLA

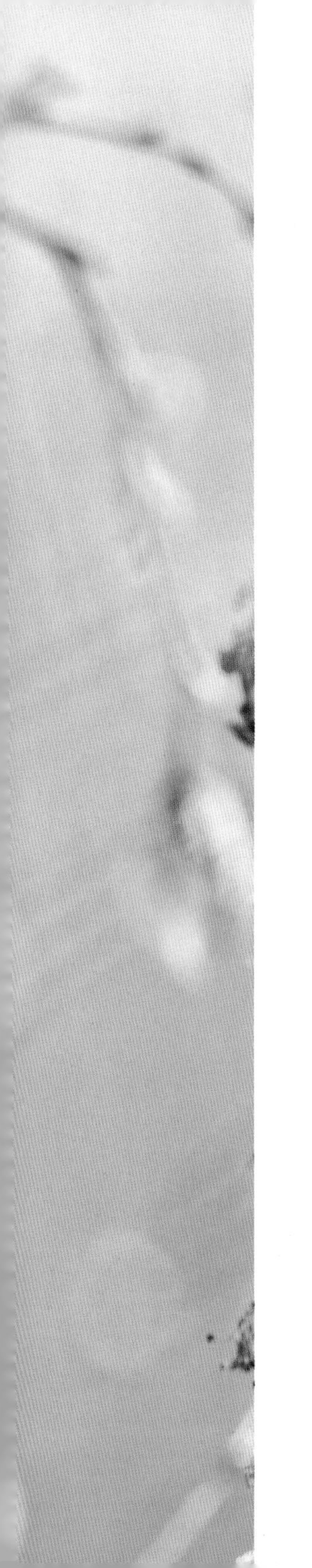

The Owl Handbook

Investigating the Lives, Habits, and Importance of These Enigmatic Birds

JOHN SHEWEY

Timber Press • Portland, Oregon

Frontispiece: Long-eared Owl

Photo and illustration credits appear on page 229.

Timber Press
Workman Publishing
Hachette Book Group, Inc.
1290 Avenue of the Americas
New York, New York 10104
timberpress.com

Timber Press is an imprint of Workman Publishing, a division of Hachette Book Group, Inc. The Timber Press name and logo are registered trademarks of Hachette Book Group, Inc.

Printed in China on responsibly sourced paper

Cover design by Sara Isassi
Text design by Lauren Michelle Smith

ISBN 978-1-64326-227-7

A catalog record for this book is available from the Library of Congress.

Contents

Introduction

Mysterious Birds of the Night

Owls intrigue us. In a way, they even look like us, with those big, bright, round eyes that stare straight ahead, seeming to penetrate our very souls. Their after-hours lifestyles have long made owls difficult to study. They remain mysterious in many ways, and scientists routinely make new discoveries about them.

Because owls are creatures of the night, at least in popular perception if not always in reality, and because so many of them share that stern, studious appearance and utter exotic and eerie calls, these iconic birds have been both vilified and lionized throughout history. Some cultures venerated owls, while others considered them abominations, even witches and evil spirits. In the Americas, among Native American cultures, owls serve a variety of mythological roles. They can be harbingers of bad news or even death, messengers from or escorts to the spirit world, warrior spirits, evil spirits, even guardians and guides; their calls can serve as warnings or bad omens. In some Native oral traditions, certain owls are feared and some even mocked. Native stories about owls are numerous and diverse, as varied as the indigenous cultures of the Americas—and given that the New World has been occupied by humans for more than 20,000 years based on the latest archeological evidence, just imagine how many tantalizing, perhaps terrifying owl tales, carried on only through oral traditions, have been lost to time.

In all cases, superstitions about owls were slow to surrender to the discoveries of science. In the Old World, by the time the printing press was ubiquitous at the end of the late Middle Ages, Renaissance thinkers were already examining many old superstitions. Sir Thomas Browne, for example, compiled a vast list of fallacies in his *Pseudodoxia Epidemica: Or, Enquiries Into Very Many Received Tenents, and Commonly Presumed Truths* (1646), including persistent old myths about owls as harbingers of doom (*pseudodoxia* means erroneous beliefs, and *epidemica* means widespread). Voices of reason continued to make inroads over the next few centuries, such as in 1874 when Alexander Young, in an article titled "Birds of Ill Omen" in the influential *Atlantic Monthly*, assured readers, "The owl has had to bear a good deal of unmerited abuse because of his nocturnal habits and unmelodious notes."

But old fears and falsehoods dissipate begrudgingly, and in some parts of the world owls are still persecuted in the name of deeply held delusions. On the Indian subcontinent, for example, the five-day Diwali—the festival of lights—is not so festive for owls. They are still illegally killed, and captured and tormented, in substantial numbers to fulfill a variety of religious and cultural beliefs.

In much of the world, such persecution, often in the name of mythology, has waned substantially, but even as superstitions about owls slowly yielded to a better understanding of these enigmatic birds, another antiquated prejudice came to the forefront. The idea that killing predators helps preserve populations of game animals for hunters dates back centuries, but in the United States this practice reached cruel extremes in the 19th and 20th centuries. As state and federal agencies began taking active roles in promoting hunting and fishing, wildlife managers, frequently influenced by private clubs and associations, embraced old ideas about controlling predators to benefit game species. The irony was hardly considered: they were killing animals so that hunters could kill other animals. Owls were on the chopping block because of the mistaken impression that they eat too many game birds.

An early naturalist, Gene Stratton-Porter (1863–1924) lent a pioneering voice advocating for ecology in general, and for her favorite birds, owls. In the mid-1920s, she wrote a series of "Tales You Won't Believe" stories for *Good Housekeeping*, one of which was titled "The Bird that Needs a Champion."

"From Time immemorial it has been the custom to consider the owl a bird of evil omen," wrote Stratton-Porter. "Pliny, Aristotle, and Aristophanes each said so. Attar, the old Persian poet, shared these prejudices. . . . I was reared on these ancient pronouncements against the owl, yet so soon as I could walk alone I constituted myself the personal champion of these birds."

Through various writings, Stratton-Porter captured the attention of at least some hunters in influential positions. At the time, owls, hawks, and numerous mammalian predators were systematically killed in the name of game management, and Stratton-Porter's advocacy for owls was repulsive to the likes of G.H. Corsan, who in the October 1923 issue of the *Game Breeder and Sportsman*, opined, "Gene Stratton Porter defends the owls and in my last game notes I mentioned this and criticized her for doing so. Now I like Nature lovers to be balanced and I cannot see anything lovely in the vicious destructive nature of the screech owl nor the great horned owls. Their most hideous noise is truly indicative of their hideous natures."

Although scientific studies were already proving that most owls prey primarily on rodents and other small mammals, owls and other raptors were still seen as vermin. Even the groundbreaking Migratory Bird Treaty Act, signed into law in 1918, offered no protection for raptors. Astonishingly, more than half a century would pass before birds of prey were afforded full protection by a 1972 amendment to the Migratory Bird Treaty Act. Today, all raptors in North America are protected by law, yet here and around the world, many owls face dire futures from habitat loss and other threats.

Some species of owls have gone extinct within the past few centuries, generally because of humans altering habitat and by the inevitable rats, cats, and dogs that accompanied seafaring Polynesian and European explorers and settlers. Many of the planet's 250-odd species of owls are in severe decline, with several on the brink of extinction.

However, extinction—oblivion—is not inevitable, even for the rarest species. Saving them requires concern, recognition, and action. But first comes understanding, the foundation of wisdom, and with this book I hope we can all learn a little more about these amazing birds that haunt the night.

1

Owls

Fun Facts and Farcical Fictions

I like their big, wise eyes, surrounded by symmetrical reflectors such as no other birds have. I like the exquisite markings and the colourings of their plumage; I am forever marveling at the velvet softness of their flight through the mystery of the night hours. I think their strange power to intensify their vision according to their requirements is miraculous. I even like their wavering courting calls in the forests of later winter.

—GENE STRATTON-PORTER, 1924

The discomfort of bitter cold was no impediment to fun in my childhood when we lived atop a hill, over a mile high, in a rural neighborhood in southeast Idaho. Unrelenting windblown snow welcomed us during our first winter on "the hill," but for us kids—my sister and brother and our friends who lived there, in houses that sat on multi-acre lots—the snow was joyous. Snow meant sledding, and tunneling, and igloo-building, and—of course—snowball fights! We hardly noticed the cold, even as our parents bundled us in bulky layers and galoshes, mittens, and knit caps.

Great Gray Owls

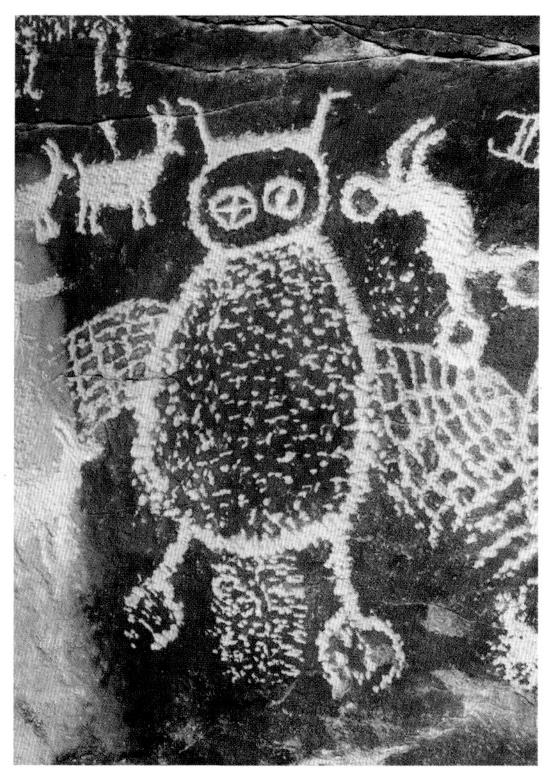

This Native American petroglyph in Utah probably depicts a Great Horned Owl.

In February of our first winter on the hill, we began hearing the deep, haunting hoots of an owl somewhere nearby. Each night, not long after the early wintertime dusk of the Northern Rockies, the hooting began. Mom told us it was a Great Horned Owl. Dad set about trying to find it.

The open landscape presented few perches for owls—scattered junipers, a grove of aspens half a mile distant, tangled alders along a small creek off to the north. But this owl was nearby. Armed with his big trusty flashlight, Dad bundled up and headed out into a steady, relentless snowstorm. He was gone mere minutes, returning to fetch all of us. We donned warm clothes and shuffled out into the snow, huddling against a bitter breeze. Dad led us quietly up to where our driveway met the gravel road and then, with the beam of his flashlight, traced the outline of a wooden power pole upward into the slanting snow.

And there he was: a Great Horned Owl perched atop the power pole, gently illuminated through the puffy snowflakes, his breast feathers dancing in the wind. He turned first toward us, then away, then back again. We could just barely make out his penetrating yellow eyes. I was hooked on owls.

Haunting the Night for Eons

In the seesaw game of predator versus prey, animals wage an eons-long battle of evolutionary one-upmanship, and deciphering why organisms adapt certain traits is a daunting task for evolutionary biologists. Owls provide an excellent example when we ask the fundamental question: why do most owls hunt in the dark hours between dusk and dawn? For an owl, an animal physiologically equipped for hunting in the dark, the cover of night proffers some advantages: lack of competition for finite prey from day-hunting species; the opportunity to hunt prey that is only active at night, using the cloak of darkness to avoid becoming prey instead of predator; and the luxury of nocturnal secrecy from small birds that—during the day—loudly raise the alarm when they happen upon an owl trying to sleep. Other reasons may also come into play. But ruling the night requires certain tools, and owls are well-adapted to the task with their superb low-light visual acuity, excellent hearing, and feathers that are structured to minimize flight noise.

Because of their nocturnal habits, owls have been associated with dark spirits, sorcery, witchcraft, and other evils since time immemorial, and their spine-tingling calls, piercing eyes, and silent flight serve to reinforce fears and superstitions. Not surprisingly, many religions tend to treat owls unkindly; in the Bible, in Leviticus, for example, the owl is an abomination, an unclean animal to be detested.

Historically, when religions gain ascendency, their burgeoning popularity tends to suppress other belief systems, and when the spread of Christianity hastened the demise of many deities from other cultures, the reputation of owls suffered. One example is the Baltic goddess Ragana, often associated with the owl, who was demoted as her followers were converted to Christianity. Once demonized she became little more than a mythical witch in league with owls, both creatures haunting the night.

Cultures the world over, across thousands of years, stigmatized the owl, and deeply held anxieties spread easily in the days before science began demystifying these enigmatic birds. In one old tale from America's colonial period, as retold in *The Animal Kingdom* (1844), a party of Scotsmen had looted the tomb of a Native American, then camped nearby for the night, but "hearing at intervals the dismal cry of the Horned Owl, endured an agony of fear, supposing it to be the moaning of the departed spirit of the Indian, whose tomb they had taken to pieces."

As was (and still is) the case on every continent, owls are treated differently by various cultural groups. Among the myriad Native American cultures, owls served many different roles in myth and folklore. A Delaware Tribal story recalls a time of food shortage that led a warrior to capture an old white owl to feed to the maiden to whom he was betrothed. The wise owl dissuaded the warrior, however, by offering to serve as the man's ever-watchful guardian. In return, the owl's life was spared, allowing him to feed his own wife and children and attend his daughter's wedding. Upon agreeing to the owl's bargain, the warrior soon found a herd of deer and was able to avert impending starvation. The old white owl performed his task as guardian thereafter, culminating in saving his adopted people from ambush by another tribe by loudly announcing their furtive approach.

Some tribes even invoked owls to frighten young children into being quiet. The Kutenai (aka Kootenay) of the Northern Rockies, for example, used the threat of kidnapping by an owl to quiet noisy or crying children, and their oral tradition includes the story of the Coyote and the Owl. In this tale, the kidnapping owl is finally outwitted by the coyote, who returns the abducted children to their parents by changing himself into a human child and crying loudly; the owl comes for him, and he allows himself to be carried off. At the owl's lair he turns himself back into his true form and kills the owl. The story varies considerably with the telling, as preserved versions attest, but in all accounts the owl is clearly a threat to children who don't obey when told to be quiet.

All in the Families

Scientists divide owls into two taxonomic families: the "true" (or "typical") owls (Strigidae), and the Barn Owls (Tytonidae). Strigidae includes more than 200 species, grouped into two dozen genera. Members of this diverse family are found everywhere on the globe except Antarctica, and their size range is remarkable. The largest owl is the Blakiston's Fish Owl, which can weigh 10 pounds and typically weighs 6 to 8 pounds—nearly as heavy as an average Bald Eagle. It's a real heavyweight compared to the aptly named Elf Owl, which is the size of a sparrow, at about 5.5 inches in length. The family name, Strigidae, recalls ancient superstitions about owls: *striges* is the plural form of the ancient Greek word *strix*, which at various times and places referred to a mythological bird of ill omen and to a witch.

Tytonidae, meanwhile, comprises about 16 species, including the familiar Barn Owl and its many similar relatives, along with the Oriental Bay-Owl and the Sri Lanka Bay-Owl. Members of the Barn Owl family are found over much of the globe—and differ in shade, markings, and size—but they share a family resemblance characterized by a heart-shaped face, long legs, and streamlined appearance. And only one is called Barn Owl. Though all are similar, *Tyto* also includes Sooty Owls, grass-owls, masked-owls, and others.

Owls are carnivorous, with sharp claws (talons) and hooked bills, like other birds of prey, such as hawks, eagles, and falcons. All these predatory birds, including owls, are called *raptors*, a word that derives from the Latin *rapere*, meaning to seize and carry off by force, but the owl's precise evolutionary path and kinship to other raptors remains under study. For a long time, scientists thought owls were most closely related to the family of birds collectively known as nightjars (Caprimulgidae), which includes such familiar species

Barn Owl

as the Eastern Whip-poor-will and Common Nighthawk. Like owls, these birds are active in the dim hours; some are nocturnal and others are crepuscular, active around dawn and/or dusk. But their relationship to owls may end there, and scientists continue to refine what we know about owls and their lineages. Owl evolutionary history, along with owl taxonomy (especially within the Barn Owl family) remains controversial and under study.

Puzzling, Perplexing Pygmies

Speaking of controversial and under study, Barn Owls and their *Tyto* kin got nuthin' on pygmy-owls. About 17 species of these little mighty mites inhabit the Americas alone—"about" because scientists aren't sure how many species and how many subspecies make up this taxonomic conundrum. Ornithologists trying to work it all out might well relate to 19th-century British author Francis Francis,

Northern Pygmy-Owl

who upon visiting Grantown-on-Spey in Scotland found himself flummoxed, saying, "Grantown is very properly so named, being in the heart of the Grant country; and every other man's name in the place—to judge by the shops, etc.—is Grant; and how the dickens they manage to indicate this Alexander Grant from that Alexander Grant, or that one from t'other, and t'other from another one, and another one from some other one, and so on, would puzzle any stranger."

Late in the 19th century, the American Ornithologists' Union (AOU) ordained one species and a subspecies for the pygmy-owls of the United States and Canada: we had *Glaucidium gnoma*, the Pygmy-Owl, and its Pacific Coast version, *G. gnoma californicum*, the Pacific Pygmy-Owl. But not everyone agreed that the Pacific Pygmy-Owl was a mere subspecies. Moreover, some researchers of the mid-20th century recognized up to seven subspecies. In the 1990s, a taxonomic revision resulted in the Northern Pygmy-Owl (*G. californicum*) and the Mountain Pygmy-Owl (*G. gnoma*). Then a few years later, the AOU (now the American Ornithological Society) decided on lumping rather than splitting, and suddenly the Northern Pygmy-Owl comprised not two, but four subspecies when *G. cobanense* (Guatemalan Pygmy-Owl) and *G. hoskinsii* (Baja Pygmy-Owl) were brought into the fold to form one big happy pygmy-owl family.

But we're nowhere near the end of this plus-size problem with pint-size pygmies. Various organizations and ornithologists abide by the idea that *G. californicum* is the Northern Pygmy-Owl, and *G. gnoma* is the Mountain Pygmy-Owl; others maintain that *G. gnoma* is the Northern Pygmy-Owl and a batch of the others are subspecies. The lab cats will eventually work it all out and tell us which pygmy-owls get to be full species, which are subspecies, and which are which. Maybe. Meantime, Cornell University's Birds of the World experts, with a firm handle on understatement, say "the situation remains unclear." And remember those dozen-plus *other* pygmy-owls of the Americas? Yeah, they're part of this taxonomic jigsaw puzzle, too, and Birds of the World admits, "Taxonomy unresolved. Phylogenetic analysis using vocalizations, morphology, and molecular data for entire *Glaucidium* genus and entire *gnoma* complex still needed."

If you're lucky enough to get a close-up view of a pygmy-owl, its adorable cuteness will probably override any concerns you have about addressing it by its proper Latin name. The songbirds and rodents these tiny raptors eat may not agree with this sentiment, but pygmy-owls are cool. And so is the fact that they are still taxonomically mysterious.

Holding Court?

We have a gaggle of geese and a covey of partridge, even a murmuration of starlings and an unkindness of ravens. These oft-farcical names for congregations of creatures derive from Middle Age English terms of venery. In those days, *venery* referred to the lexicon of hunting, and as journalist Nelson Handel explained in a story for *Los Angeles* magazine, "the first outbreak of venereal terms appeared in the so-called Books of Courtesy, social primers that strove to codify manners at court." Today we associate the word "court" with legal proceedings, but in those bygone days of English high society, the books of courtesy taught people of privilege how to act and speak in polite company of their peers and peeresses—people

A Parliament of Owls, by Terry Wing.

holding inherited or honorary titles. Books of courtesy, which also appeared in France, Germany, and other medieval feudal societies, instructed courtiers in etiquette, morals, social graces, and more.

Hunting was a major pastime among the elite of the Middle Ages, and subject to its own protocols and decorum, so not surprisingly, the art of the hunt—pursuing stags (deer), fox, or hares (usually with hounds), hawking (hunting hares or game birds with falcons), and fowling (hunting game birds)—was addressed in some books of courtesy. The first manners book in England was the *Book of the Civilized Man* by Daniel of Beccles, which appeared early in the 13th century, but dozens more followed, the best-known being *The Book of Saint Albans* (1486). This book popularized terms of aggregations (for animals and people) owing to an exhaustive list therein. Most of these archaic, obscure, and enigmatic terms are little used now (and probably then, too), popping up only in trite Internet trivia lists and the like. Some remain common in the lexicon of both wild and domestic animals, such as a pride of lions, a flock of sheep, and a swarm of bees. But dozens upon dozens of others from the list are cryptic, at least to our 21st-century viewpoint. For example, why a sleuth of bears and an ostentation of peacocks?

Owls, however, don't appear in "Companynys of beestys and fowlys" from *The Book of Saint Albans*, an omission that is unsurprising given that owls are solitary birds. A few species, such as the Short-eared Owl and Long-eared Owl, occasionally roost communally in areas with a dearth of good roosting sites, but usually only owl families—fledglings and parents—form congregations. And yet, modern-day trivia lists of funny names for congregations of a single species almost always include the perplexing term "a parliament of owls." The oft-given rationalization for this odd idiom is that it derives from the association of owls and wisdom, which has persisted for centuries and is avouched by the still-common phrase "wise old owl."

Yet "parliament of owls" may not be a Middle Ages term at all: in the 1950s, C.S. Lewis coined the quirky phrase in his classic and popular series *The Chronicles of Narnia*, in which a council of owls convenes at night to discuss goings-on in Narnia. Lewis's masterpiece was so influential that "parliament of owls" quickly took hold as the de facto designation for a group of owls (regardless of whether they form groups). If the term did predate Lewis, then it likely derives from mistaken pronunciation of the homophonous title of a 1380s Geoffrey Chaucer poem that inspired Lewis's council of Narnian owls: *Parliament of Fowls.*

Of Owls and Gods

"That Owls should be in such esteem at learned Athens, as to stamp their pictures on their coyn, to me is strange," opined Alexander Ross in 1647. Ross, labeled "the vigilant watchdog of conservatism and orthodoxy" by 20th-century writer Richard S. Westfall, still clung to the ages-old loathing of owls, born of misunderstanding.

But more than 2,000 years before Ross wrote *Mystagogus Poeticus, or The Muses' Interpreter*, the Greeks had indeed depicted owls on coins honoring the sagacity of Athena, their goddess of wisdom and war. The Little Owl, a widespread old-world species, was idolized, especially in Athens, and legend says that on the eve of the Battle of Marathon in 490 BCE, the Owl of Athena appeared in the sky, a harbinger of victory for the Athenian armies under Miltiades against the invading Persian forces. A decade later, according to Greek mythology, the sight (or call) of Athena's owl likewise foretold

The Greek goddess Athena on terra-cotta, circa 485 BCE, held by the Metropolitan Museum of Art

The Little Owl, a widespread species in Europe, Asia, Africa, and the Middle East, was revered by the ancient Romans and Greeks. This Greek coin depicts the Little Owl, which was associated with the goddess Athena.

ill fate for the Persian fleet at the Battle of Salamis. Some decades later, the Greek coin (called a *tetradrachm*) bearing a likeness of Athena on one side and her owl on the other was in widespread circulation.

And despite Roman aversion to owls—at least the large species, such as the Eurasian Eagle-Owl—they, like the ancient Greeks, revered owls associated with Athena's Roman incarnation, Minerva. The Little Owl constantly accompanies her in Roman legend, Minerva took pity on the shunned Nyctimene, daughter of King Epopeus, and transformed her into an owl. When science began to supersede superstition, the Little Owl earned the Latin name *Strix noctua* in 1769. But in 1822, German zoologist Friedrich Boie successfully changed the name to *Athene noctua* to honor the Greek goddess.

Virtually every religion evokes the owl in some form, either vile or virtuous, or often both, depending on context. In Hinduism, for example, owls are generally reviled, but many Hindus in Bengal revere the Barn Owl

A mother Burrowing Owl delivers an insect to a hungry chick.

for accompanying both Lakshmi, the goddess of wealth and prosperity, and her opposite form, Alakshmi, the goddess of inauspiciousness. Named Uluka, the owl, always white in its depictions, becomes Lakshmi's *vahana* (vehicle) when the prosperity she provides to humans is turned to *adharmic* (immoral) vices. Uluka also serves as a symbol reminding people to use wealth wisely and justly.

The Ole Honey Bucket Trick

The Burrowing Owl is unique, the only owl that lives underground. Across most of their range in the Americas, these fascinating birds commandeer burrows made by ground squirrels and other animals, make improvements as needed by scratching and digging with their feet and bills, and use the burrows for nesting, rearing young, roosting, and life in general; in some places, especially Florida and the Caribbean islands, they typically dig their own burrows. The iconic photo of this species depicts the owls standing upright on their long legs near their burrow. They eat a variety of prey, including small mammals, small birds, reptiles, and—significantly—insects.

Beetles are especially yummy to these little owls, and they use a clever trick to make beetle-eating easier. In many places, Burrowing Owls collect scraps of mammal dung and scatter it around their burrows, attracting dung beetles and other beetles, a behavior that puzzled scientists for a long time. Finally in the early 2000s, scientists from the University of Florida in Gainesville discovered that attracting beetles was precisely the point of stockpiling poop. Zoologist and lead author of the study, Douglas J. Levey, likened the strategy to an angler casting a line into the water to lure fish.

A form of tool-using by an animal, this evolved predatory strategy was first posited as a means of helping protect nestling owls by masking their scent, as poo is apt to do, but experiments soon disproved that theory. So, the researchers designed additional experiments to test the fishing-for-a-feast idea and found that the owls at the study colonies ate 10 times as many dung beetles when their burrows were surrounded by cow manure.

Burrowing Owl

Hand in the Cookie Jar

Nesting underground has its potential pitfalls. Burrowing Owls, especially juveniles, are squarely in the sights of the same cunning creatures that target the rodents that dig burrows, and a baby owl whose parents commandeered the burrow is a fine snack for a hungry predator. As Michael Aranda of SciShow says, "A burrow is kind of like a cookie jar—an easy-to-open container with a tasty treat at the bottom."

Badgers, coyotes, bobcats, and foxes are happy to dine on ground squirrels or young owls, given the chance, so how do Burrowing Owls avoid unwanted dinner guests that might make them the dinner? Simple: sound like something a predator wants to avoid, in this case a rattlesnake. Burrowing Owls have evolved a drawn-out hissing call that sounds remarkably like a rattling rattlesnake—a creature that most mammalian predators want no part of, especially if the snake has taken shelter in a hole in the ground. Any predator that sticks a foot or a face in that hole might just meet the business end of a venomous snake, so Burrowing Owls use a snakelike sound to fool would-be party-spoilers.

In the 1980s, researchers published the results of a study that demonstrated that the

The Long-eared Owl is one of the many species of owls with prominent ear tufts.

Burrowing Owl's hiss is a case of acoustic Batesian mimicry. *Acoustic*, of course, refers to sound; the rest of that mouthful—*Batesian mimicry*—refers to a defense mechanism in which a harmless species evolves the ability to imitate warning signals of a dangerous species as a means of fooling potential predators. The name derives from English naturalist Henry Walter Bates, who studied such mimicry in Brazilian butterflies.

Fall on Deaf Ears

Does a Long-eared Owl have long ears? Does a Great Horned Owl have horns? At first glance it might seem so. These are just two of the many owl species with tufts of feathers on their heads that look like upright ears, or horns. But these adornments serve no sensory purpose—they are not ears at all (or horns). So why do dozens of owl species have ear tufts? That question continues to inspire researchers, who have posited several theories.

One line of reasoning is that an owl's ear tufts may help the bird communicate with members of its own species and with other species—perhaps even functioning to make the bird appear more menacing to potential predators. Owls can raise and lower their ear tufts, and when fully raised, these tufts are distinct and well defined in most "eared" species. Pairs or families of owls may use ear angles (lowered or raised) to communicate and keep track of one another. They may also serve to make the bird look bigger and more fearsome to predators, but if that were the case, why don't all or even most owls have ear tufts?

A more widely accepted theory is that ear tufts serve to help camouflage a roosting owl.

The reasoning is that a smooth, rounded head stands out more amid tree trunks and branches, whereas that set of feather tufts aids in making the owl's head look more like a broken limb top. The camouflage theory would predict that owls with ear tufts are forest/shrub inhabitants (owls that roost in trees) rather than open-country owls, and that they are nocturnal, because diurnal owls that roost at night enjoy the cover of darkness to conceal them when they are roosting. Sure enough, none of the diurnal and mostly-diurnal owls have well-developed ear tufts, and all the chiefly nocturnal owls with ear tufts are denizens of wooded habitats.

Problem solved?

Not so fast. If better camouflage for forest-dwelling nocturnal owls explains ear tufts, then why don't *all* forest-dwelling nocturnal owls have conspicuous ear tufts? In fact, a majority do not. That's why researcher Michael Perrone Jr., after examining various theories on the matter of owl ear tufts, concluded that "the only proposal concerning the function of ear tufts that finds support from my tests is that of camouflage. However, the camouflage hypothesis does not fully explain the occurrence of tufts either, since more than half of the nocturnal, forest-dwelling owls lack tufts, despite the advantage these structures presumably confer. A firmer conclusion awaits field observations of perching and roosting behavior of owls, including site preferences and postures."

An owl's facial disc—the rim of narrow feathers that frames the bird's face, such as on this Ural Owl (a Eurasian species)—acts like a radar dish that helps channel sound to the ears.

Hear Ye, Hear Ye

An owl's real ears are slits hidden by feathers, located on the sides of the head behind the eyes, and owls have larger eardrums relative to body size than most birds. Owl hearing is so efficient at zeroing in on sounds made by prey that various owl species have been observed pouncing into the snow and emerging with a rodent that the bird could not possibly have seen.

Additionally, an owl's face is structured to enhance the bird's hearing. Most owls have a well-defined facial disc—the narrow rim of small but prominent feathers that form a border around the face, making the entire face something akin to a radar dish. These specially-structured feathers help channel sound to the ears.

Moreover, in a remarkable adaptation for hunting by sound, many owl species have ears that are asymmetrical in their placement—the right ear is slightly higher on the head than the left ear, and the two are also slightly misaligned on the vertical plane. This means that a sound, such as a vole scurrying in the grass, reaches one ear ever so slightly before the other, which helps the bird pinpoint the location where the sound emanates from—like a sophisticated form of triangulation. Once the sound registers in the owl's neural system, the bird can then use interaural time difference (ITD) and interaural level difference (ILD) to pinpoint both the location and elevation of the source of the sound—typically the animal they are hunting.

Combined with exceptional vision, an owl's ability to locate prey by sound makes them near-perfect crepuscular killing machines and formidable nocturnal hunters. Not only can they use sound to locate prey hidden by snow or vegetation, but they can zero in on these slight noises so effectively that the unwitting victim has little chance of escape.

We Need You, Woody!

"Eco-engineers" and "keystones species"—those are two epithets sometimes used to describe woodpeckers because these avian excavators are critical cogs in the ecosystem. Woodpeckers chisel out holes in trees to build their nesting cavities and to hunt for food, and in so doing inadvertently create nesting and hiding habitat for many other creatures, including numerous species of owls. For example, all but one of the small owl species found in North America nest in tree cavities (the exception being the Burrowing Owl), and several medium-size species will nest in woodpecker holes of the right size and location. Small owls also routinely use woodpecker holes for roosting. This critical connection between woodpeckers and cavity-nesting owls (and other cavity-nesting birds) is called a symbiotic relationship: without the woodpeckers, owls would suffer a severe shortage of nesting cavities. That's why trees, and more to the point—forests—are so critical to owls worldwide.

The largest North American woodpecker is the crow-size Pileated Woodpecker, found throughout the eastern United States, most of Canada, the Pacific Northwest, and much of California. These giants of the woodpecker world are especially fond of carpenter ants and wood-boring beetles and hunt for them by literally tearing trees apart, especially dead, decaying trees (where carpenter ants thrive). For nesting, Pileated Woodpeckers excavate large holes, usually in dead or dying trees, which often have hollow or rotting cores that make the bird's job easier. They also excavate

The Pileated Woodpecker is a critical habitat creator.

A Red-bellied Woodpecker checks out an Eastern Screech-Owl. Small owls often nest in cavities excavated by woodpeckers.

roosting sites, often in live but decaying trees, and they chisel out multiple entrance/exit holes in nesting and roosting cavities. Woodpecker woodworking on such a massive scale means lots of cavities for owls.

Sporting a prominent red crest and flashy black-and-white plumage, the Pileated Woodpecker might seem like the obvious inspiration for Woody Woodpecker of cartoon fame, but legend says that an altogether different species—the Acorn Woodpecker of the Far West—led to animator/producer Walter Lantz creating his iconic laughing woodpecker character.

In addition to the Pileated Woodpecker, North America's other large woodpeckers are equally critical to small owls. Among the most widespread, the Northern Flicker creates holes and cavities in woodlands throughout the continent, as does the somewhat smaller Hairy Woodpecker. The Northern Flicker's Sonoran Desert cousin, the Gilded Flicker, is especially important for providing nest holes for Elf Owls, often in saguaro cacti. In fact, in the Desert Southwest, several small species of owls rely heavily on cavities excavated by a variety of woodpeckers found only in that region, including the Golden-fronted Woodpecker, Gila Woodpecker, and Arizona Woodpecker. In the East, alongside the wide-ranging Hairy Woodpecker, colorful Red-headed Woodpeckers and Red-bellied Woodpeckers provide cavities repurposed by small owls.

The Eyes Have It

You won't prevail in a stare-down with an owl. They seem to see *through* you, right into your very soul. A face-to-face encounter with one of the big, yellow-eyed owls—a Great Horned, Great Gray, or Snowy Owl—is unnerving to some, deeply moving to others. "A volume could be written on the eyes of an owl," said the duly impressed American author and naturalist Gene Stratton-Porter. She was understandably mesmerized by the unflinching wide-eyed gaze of her favorite type of bird and keen to appreciate the things that make owl eyes so special.

Most owls hunt in dim light or darkness, using their specialized ears and excellent night vision to zero in on prey. Some species are equally adept at hunting in broad daylight, when both superb hearing and exceptional vision make them efficient predators. Most birds and mammals have rounded eyes, but an owl's eyes are tubular and locked in place by a sclerotic ring, which is a series of platelike bones. Unlike our eyes, an owl's eyes are fixed; the bird cannot shift its gaze side to side or up and down, which largely explains that penetrating gaze. Instead, because they have twice as many cervical (neck) vertebrae as we do—14 instead of 7—owls can turn their head 270 degrees and almost completely upside down. With eyes on the front of their faces, owls have about 70 degrees of binocular vision, which allows them to see in three dimensions, able to judge distance and size like we do. In some species of owl, the eyes account for some 5 percent of body weight and 50 percent of the skull area.

Great Horned Owl

Owl eyes are packed with rod cells—the kind of photoreceptor cells that allow for night vision. They have far fewer cone cells, which provide for discernment of colors, so where we see lots of bright shades thanks to more cone cells in our eyes, owls tend more toward seeing the world in gray tones. However, researchers continue to study how owls see their world, and new details emerge frequently. One thing we know for certain is that owls see a lot better in the dark than we do because their comparatively huge eyes are designed to gather far more light than ours.

Many owls have dark-colored irises—their entire eye appears black, although the iris is actually dark brown. But other species have bright yellow eyes (or bright orange in some parts of the world). Some of the most familiar North American owls, such as the Great Horned Owl and Eastern Screech-Owl, have piercing yellow eyes. For a long time, owl experts thought that yellow/orange eyes occurred in owls that are active during daylight, and dark eyes occurred in owls active at night. But in 2018, a team of Spanish researchers published the results of a comprehensive analysis of owl eye colors and discovered that iris color—dark versus yellow—is not as predictive of nocturnal versus diurnal activity as once thought. "The proportion of dark-eyed owl species is higher among strictly nocturnal owls than among diurnal ones," they discovered, but the correlation is not particularly strong: lots of nocturnal owls have bright-colored irises and about three dozen diurnal/crepuscular owl species have dark irises.

STAMP OF APPROVAL

Rare birds, endangered birds, and birds we find fascinating inspire artists the world over, and our feathered friends have appeared on coins since the heyday of Greek and Roman ascendancy, testing the artistic skills of minters for centuries. But postage stamps are a much more recent invention. The first postage stamp was issued in England in 1840, and the idea was met with great enthusiasm—by 1860, 90 nations had issued postage stamps. Early in the proceedings, some nations began depicting birds on their stamps, such as the 1869 United States stamp showing an eagle with wings spread. By now, well over 10,000 different bird-theme stamps have been issued worldwide, and owls have been prominent for decades, much to the delight of many philatelists (stamp collectors). Brilliant, intricate artwork adorns many such stamps, and owls of all genera and dozens upon dozens of species have enjoyed their time in the stamp spotlight.

Owls have graced postage stamps issued by countries from around the globe. This Hungarian stamp from 1962 depicts a Eurasian Eagle-Owl.

Silent Night, Deadly Night

Beware, little rodents: hawks that soar the daytime skies are fearsome enough, but at least you can see them. And as a last resort you can hear them. But not so the owls, come nightfall. In the darkness, owls are hard to see and, to make matters worse, they fly in silence, thanks largely to specially adapted feathers. Nocturnal hunting rewards the quiet and stealthy owl because its prey tends to have excellent hearing. Owls that hunt in the dark have evolved flight feathers with comblike leading edges, velvety texture, and wispy trailing edges (called fringes); these features work in unison to reduce turbulence and dampen sound produced by flapping and gliding.

Moreover, silent flight not only makes flying owls hard to hear but also helps them use their own superb hearing to zero in on prey, unimpeded by noise from wing flaps. Scientists call this acoustic tracking. Night-hunting owls also tend to have larger flight feathers and wings relative to body size than most birds, an adaptation that allows them to fly

Thanks to unique feather attributes, owls are the quietest fliers in the bird world.

slowly to further reduce noise. Many owls, such as the Barn Owl and Short-eared Owl, can fly so slowly that their flight pattern is often described as mothlike.

Testifying to the importance of silent flight in nocturnal (and crepuscular) owls, those species that primarily hunt by day, or which hunt aquatic creatures and other prey not imbued with the acute hearing of small mammals, mostly lack these silent-flight adaptations. For example, the Northern Hawk Owl, a circumpolar species at northern latitudes, hunts by day and lacks the feather adaptations for silent flight. Likewise, the three species of fishing-owls of Africa are also comparatively noisy fliers because they specialize in aquatic prey.

This Italian red wine is called Cala Civetta, which translates to "Owl Cove," a secluded beach along Italy's Mediterranean coast.

Hair of the Dog . . . or Egg of the Owl?

Gaius Plinius Secundus (23/24–79 CE), better known to us as Pliny the Elder, was a Roman naturalist and philosopher (and military commander) who is most famous for his sprawling, deeply detailed, 10-volume *Natural History*. Essentially an encyclopedia of both the natural and the human worlds and their many intersections, *Natural History* is a vast compilation of facts, fancies, fictions, and ideas largely from other ancient authors.

In Pliny's Roman society, large owls were generally to be feared even while Minerva's Little Owl was venerated. British raptor expert Robert D. Frost explains that "the Romans saw owls as omens of impending disaster. Hearing the hoot of an owl indicated an imminent death, it is thought that the deaths of many famous Romans was predicted by the hoot of an owl, including Julius Caesar, Augustus & Agrippa. While the Greeks believed that sight of an owl predicted victory for their armies, the Romans saw it as a sign of defeat. They believed that a dream of an owl could be an omen of shipwreck for sailors & of being robbed. To ward off the evil caused by an owl, it was believed that the offending owl should be killed & nailed to the door of the affected house."

The eagle-owl, especially, was so feared that when one of these huge, horned birds was seen in Rome during the convening of the consulship of Sextus Palpellius Hister and Lucius Pedanius, the entire city was purified. In keeping with the times, Pliny related some of the old superstitions about owls. He called the eagle-owl a "funereal bird" that was "regarded as an extremely bad omen, especially in public auspices." However, Pliny the observer also relates, "I know several cases of its having perched on the houses of private persons without fatal consequences."

Pliny also recorded countless remedies for multitudes of ailments and disagreeable conditions. He was skeptical of some and disdainful of others, but neutrally informative on many. Evidently, he never tried an old remedy for drunkenness: "The eggs of an owlet," he explained, "administered to drunkards three days in wine, are productive of a distaste for that liquor."

Crack a couple hen eggs, let alone owl eggs (which are fully protected by law), into my glass for three days running and I might very well lose my appetite for wine.

Noncommittal on the idea that owl eggs might encourage sobriety, Pliny is downright skeptical of another old tale: "They [magicians] describe, too, certain remedies made from the egg of this bird for the hair. But who, pray, has ever had the opportunity of seeing the egg of a horned owl, considering that it is so highly ominous to see the bird itself? And then besides, who has ever thought proper to make the experiment, and upon his hair more particularly? In addition to all this, the magicians go so far as to engage to make the hair curl by using the blood of the young of the horned owl."

Dem Bones

Owls are among the many different types of birds that regurgitate undigestible leftovers from their prey—bones, fur, claws, teeth, feathers, insect exoskeletons, and more—in the form of compact pellets that often accumulate below or around favorite roosts and nesting sites. These pellets form in a bird's gizzard—sort of an internal trash compactor—within 10 hours of the meal. Owl pellets are usually oblong, and large owls, such as Great Horned and Great Gray Owls, produce large pellets—up to 4 inches long. The smaller the owl, the smaller the pellet. Lacking acidic digestive fluids, owls produce pellets with well-preserved remnants, allowing scientists to collect and analyze them to determine owl diets, and also to gain insights into what kinds of animals live within the owl's habitat.

Some owl pellets have even turned out to contain bird leg bands (the little metal sleeves that ornithologists place on captured and released birds of all kinds to learn more about

Owls can't digest bones and other hard body parts of their prey, so they regurgitate them as oblong pellets.

Researchers dissect owl pellets to learn more about owls and the places they live.

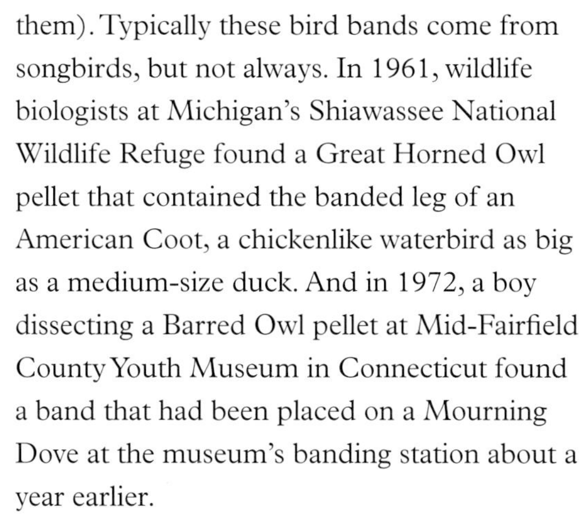

them). Typically these bird bands come from songbirds, but not always. In 1961, wildlife biologists at Michigan's Shiawassee National Wildlife Refuge found a Great Horned Owl pellet that contained the banded leg of an American Coot, a chickenlike waterbird as big as a medium-size duck. And in 1972, a boy dissecting a Barred Owl pellet at Mid-Fairfield County Youth Museum in Connecticut found a band that had been placed on a Mourning Dove at the museum's banding station about a year earlier.

Budding scientists can buy commercially sterilized pellets to dissect, or you can collect your own. However, pellets can retain rodent viruses and bacteria, so they should be collected with gloves and bagged, then sterilized by heat treatment. The International Owl Center (IOC) explains that pellets "should be wrapped in foil and baked in an oven to kill bacteria (40 minutes at 325°F according to some research, or 4 hours at 250°F according to Carolina Biological Supply's methods)." Dissecting owl pellets is a fun and educational family or school activity, and the IOC (based in Minnesota) even sells inexpensive owl pellet dissection kits that come with everything you need.

Turning Heads

Most birds can rotate their heads 180 degrees side to side, but owls take turning heads to extremes . . . or at least to another 90 degrees. Owls can rotate their heads 270 degrees, which comes in handy because, like many birds, they can't move their robust eyes. So, thanks to special anatomical adaptations, owls can literally see behind themselves. The eerie sight of an owl facing away yet staring at you with big, piercing

Owls, such as this Great Gray Owl, can turn their heads almost all the way rearward.

The effect is even spookier when the owl is facing you, but looking the other way, appearing like a headless owl!

eyes, then swiveling all the way around to look at you again from the other side gave rise to the long-persistent myth that these birds, both revered and feared throughout human history, could spin their heads all the way around like a feathered version of Linda Blair's famous *Exorcist* character. In fact, according to *The Encyclopedia of Superstitions* (Richard Webster), an old English belief says that "If you walk around a tree containing an owl, it will turn and watch you. If you keep on circling the tree, the owl will eventually [wring] its own neck."

Moreover, owls can turn their heads almost upside down without moving their bodies. They have extremely flexible necks thanks to twice as many cervical vertebrae as we have: 14 instead of our 7. And owls have specially adapted arterial structures that assure no interruption in the supply of oxygen-rich blood to the brain during all that head twisting and turning . . . and bobbing. If you ever see an owl bobbing its head, maybe canting to the left or right as well, the bird is trying to get a better look at you or something else that has caught its attention. All that head movement mitigates the fact that an owl can't shift its eyes up, down, or sideways like we can, so the bird needs to move its head to focus on objects and to help it judge distance and size.

Head-bobbing may also serve owls in other ways, such as helping them use their incredibly sensitive ears to triangulate in on prey, somewhat like we sometimes turn our heads one way or the other to hear distant or soft sounds. Owls may also bob their heads to communicate with mates, young, others of their species, or even other species. They often do a lot of head-bobbing when they vocalize with mates and family members. But accurately judging distance from and location of objects seems to be the primary purpose of head-bobbing. Scientists call this kind of location-fixing motion parallax, which is a change in position caused by the movement of the viewer. Owls (and many other birds) bob their heads to gain a slightly different view of an object, thereby helping them pinpoint its distance and precise location.

Owls have deadly sharp talons—these daggers belong to a Tawny Owl, a widespread species in Europe.

Get a Grip

Owls have extraordinarily powerful feet; their four digits (toes) are tipped with long, stout, sharp talons. But unlike most birds, which have three forward-facing toes and one rearward-facing toe, owls are *zygodactyl*, meaning their feet have two toes that face forward and two that face rearward. They share this trait with most woodpeckers, parrots, and a few other birds, but unlike most other zygodactyl birds, owls can pivot the outer toe forward as needed to help them grip prey and perch on branches or help them walk or climb (so perhaps *semizygodactyl* is more accurate). With two talon-tipped toes pointing each direction, an owl's grasp is an inescapable death grip for whatever animal is unfortunate enough to be captured by the silent assailant.

Thieves in the Night

Owlers go owling. In other words, owlers are people who enjoy hearing and seeing owls, and when people go afield to look or listen for owls, they are owling.

But those two terms—*owlers* and *owling*—had very different meanings beginning in the late Middle Ages in England. By the 1600s, the word owler had been coined as a pejorative term for a smuggler, specifically a smuggler of wool and sheep. Illegal exportation of wool and sheep from the British Isles to Continental Europe and other places became increasingly lucrative as a succession of English monarchs each imposed duties on wool imports, beginning with King Edward I in 1275. But wool smuggling found its stride

The Smugglers—a depiction of owlers (wool smugglers) in England, from *Picturesque Chapters in the Story of an Ancient Craft* by Charles George Harper, 1909.

when King Edward III implemented protectionist policies designed to aid the English woolen cloth industry in the 14th century. Taxation on wool exports severely burdened legal exportation and created a thriving black market. His successors to the crown continued apace and by the reign of Queen Elizabeth I in the latter half of the 1500s, sheep smugglers faced harsh penalties if caught: for the first offense, a smuggler would not only forfeit the goods but also be imprisoned for a year, and upon completion of that sentence have his left hand cut off in a public market at the busiest time of day, with the severed hand then nailed up where all could see. A second offense met with execution. In 1660, all export of wool was forbidden, and by then the term owler was in vogue.

Sneaking wool out of England became so potentially profitable that smuggling became big business. Historian Richard Platt, author of *Smuggling in the British Isles*, explained, "The centre of the trade was Kent where the wool exporters were known as 'owlers.' As restrictive laws strangled the owlers' trade, they became progressively bolder, and pooled their resources, on the grounds that there's safety in numbers. Soon an owling venture involved hundreds of armed men."

Presumably the terms owling and owler entered standard parlance as a reference to smugglers carrying out their misdeeds under cover of night. Their vessels were called owling boats. By the middle of the 20th century, with English smugglers having long since shifted from wool to other profitable products, the word owling reemerged from a long hiatus but with an entirely different meaning. Nowadays, owling is routinely understood to describe the birding pursuit of seeking to see and hear owls.

It's an Owl Eat Owl World Out There

Throughout most of their range, American Barn Owls are largely nocturnal, sometimes crepuscular. Even at night, they risk predation by larger, more powerful rivals, particularly the Great Horned Owl. A copse of hardwoods I visit often in eastern Oregon is a roosting site for both Barn Owls and Great Horned Owls, but they are never present at the same time—the Barn Owls vacate the premises when a deadly Great Horned Owl resides within the trees.

Great Horned Owls, in fact, are notorious for killing other owls. They are the only major predator of the widespread Barred Owl. Like those Barn Owls in the grove of trees in Oregon, Barred Owls tend to vacate the premises when Great Horned Owls are about, and I suspect they learn to keep the noise down, too: in my rural neighborhood, Barred Owls that vocalize often in the short term tend to become mostly silent very soon—either that, or they've already been killed by their larger and more ferocious brethren. More frightening still (for owls) is the account of a Great Horned Owl killing and flying off with an adult Great Gray Owl, a story told to me in detail by a reputable source.

And daytime fliers are hardly safe: I once saw a Great Horned Owl kill a Northern Harrier (a species of hawk) that had stayed out just a little too close to dusk. I was driving a 2-mile-long straightaway near home that I affectionately dubbed "hawk alley" because of the preponderance of hawks that hunt rodents in the adjacent agricultural fields; as the gloaming approached I saw a harrier coursing low over a strawberry field when a Great Horned Owl suddenly flew from a hazelnut tree grove on the opposite side of the road, quickly gained speed, and pounced on the harrier just a foot or two off the ground, dispatching it cleanly. The owl stood atop its kill as I slowed, but I continued on my way, knowing the bird likely had nestlings in need of food.

For many species of birds, a Great Horned Owl is not to be trifled with, but other owls are equally capable of killing their kin. In fact, Barred Owls are major culprits in the decline of the already-threatened Northern Spotted Owl—the bigger, more aggressive Barred Owls displace, hybridize with, and even kill Spotted Owls. And Barred Owls don't limit their aggressions to their Northwest cousins; they will kill any smaller owl and even their own kind. In 2015, a couple in Vermont watched a Barred Owl kill and fly away with another Barred Owl, and similar anecdotal accounts surface regularly.

Taking advantage of owl-versus-owl predation, ornithologist Vincenzo Penteriani used a taxidermy-mount Tawny Owl to capture Eurasian Eagle-Owls so they could be radio-tagged—and he captured this dramatic photo in the process.

Moreover, owls not only eat other raptors, but also kill them to reduce competition for the same preferred prey, such as rodents. This is called intraguild predation. In the Old World, the Eurasian Eagle-Owl is the cruiserweight counterpart to the west's Great Horned Owl, and Spanish researchers trying different methods of capturing eagle-owls to fit them with radio tags discovered that prompting owl-on-owl intraguild predation was highly effective. Using a taxidermy-mounted Tawny Owl as bait, scientist and author Vincenzo Penteriani succeeded in luring eagle-owls into range for capture, tagging, and release, and demonstrated the routine nature of such owl-on-owl predation.

The prize for the ultimate incident of owl versus owl, however, was reported by Edward Howe Forbush in *Birds of Massachusetts and Other New England States* (1925): "I once found in the stomach and gullet of a Barred Owl the greater part of a Long-eared Owl, while in the stomach of the latter were some remains of a Screech Owl."

But Turnabout is Fair Play

Some owl species often prey on birds, while others take birds opportunistically—and owl-on-owl predation is an ever-present threat, especially where deadly Great Horned Owls and Barred Owls haunt the dim hours. But owls need to be extra vigilant during the day, which is why most species hide in well-concealed roost sites between dawn and dusk. If they venture forth too early in the evening or stay out hunting a little too deep into the morning, the dangers mount, as diurnal winged assassins begin their own hunting forays.

Cooper's Hawks are among the many raptors that will prey on small and medium-size owls given the chance.

Years ago, I made regular autumn fishing trips to a coastal estuary in Oregon, where an anadromous form of cutthroat trout prowled the brackish estuary. Anadromous fish live part of their lives in freshwater and part in saltwater. These native cutthroats typically fed voraciously in fertile saltwater during summer, then headed back to freshwater in the fall, ahead of their late-winter spawning season. I enjoyed fly-fishing for them by hiking along the tidal creeks draining the grassy estuary, and abundant wildlife redoubled the fun of these expeditions. For a while I was puzzled about the remains of Short-eared Owls I would find a few times each fall. But the cause of their demise eventually came to light: one October morning, just after sunrise, I watched a Peregrine Falcon kill one of the owls in midair. The lightning aerial attack, which left a cloud of feathers drifting slowly in the breeze, left no doubt about the owl's fate.

The falcon, with the slightly smaller owl in its talons, disappeared into the marsh grass, presumably to begin the feast.

Peregrine Falcons are swift, fearsome predators. When they *stoop* (dive), they can attain speeds of more than 180 miles per hour, making them the world's fastest animal. But for day-hunting owls, Peregrines may be the fastest menace, yet hardly the only one. Accipiters are hawks that specialize in hunting other birds, usually by remarkably swift and agile flight, often in heavy cover—the same forests and woodlands where owls hide during the day. In North America, accipiters come in three sizes: small, medium, and large. Or if you happen to be a small owl, dangerous, more dangerous, and flat-out deadly. The Sharp-shinned Hawk is the smallest, about the same size as the familiar American Kestrel (the smallest North American falcon); the nearly identical Cooper's Hawk averages about the same size as a crow; and the Northern Goshawk can be nearly as large as a Red-tailed Hawk. Any of these hawks (along with various others) will opportunistically take an owl given the chance, and fledgling juvenile owls are especially vulnerable.

Of course, leave it to the formidable Great Horned Owl to turn the tables on turnabout: a video camera trained on a Peregrine Falcon nest box installed and monitored by the Raptor Resource Project in Minnesota once captured a Great Horned Owl swooping in and pouncing on a falcon chick at night. The female falcon fought the owl tenaciously and eventually drove it away, but the chick had perished.

Fight for Your Rights!

Not all raptor-versus-raptor fights are about one bird trying to eat the other. Birds of prey often come to blows over incursions by another raptor. Owls, hawks, falcons, eagles—most all of them will defend their nests, their captured prey, and even their favorite hunting grounds against other winged marauders. Whether protecting their chicks from predation or defending a favorite hunting territory, raptors are formidable foes, particularly if the two combatants are similar in size. However, even the smallest birds of prey will mob much larger raptors, especially to protect nests and nesting areas.

Short-eared Owls and Northern Harriers (a type of hawk), for example, are not only about the same size (harriers are a bit heavier), but they hunt the same prey and live in the same places. Both species search for rodents in open country, and their hunting hours overlap. The harrier is largely diurnal, and the Short-eared Owl often hunts in the wee hours of morning and the waning hours of evening, but also at night. However, harriers can be crepuscular (active around dawn and dusk), and Short-eared Owls often hunt in daylight—the two birds are bound to come into conflict.

But the owls often get the short end of the stick in this relationship because Northern Harriers are kleptos—in this case, kleptoparasites. That means they readily steal prey from other birds, primarily Short-eared Owls.

Similarly, Red-tailed Hawks and Great Horned Owls sometimes come to fisticuffs because they occupy the same niche, with the hawk being a daytime hunter and the owl taking over at dusk. Plus, Great Horned Owls don't build nests but instead frequently use stick nests built by other birds, including Red-tailed Hawks. Should an owl decide to expropriate a hawk's nest that the hawks want to continue using, you can imagine the potential for conflict. Similar antagonistic relationships occur between Barred Owls and Red-shouldered Hawks.

Owl-versus-hawk fisticuffs come in many forms. When the heavyweights duke it out, aerial acrobatics can be dramatic, such as when this Red-shouldered Hawk ventured too close to a Barred Owl nest in Florida, with the ensuing confrontation captured by photographer Marina Scarr.

Mob Mentality

Birds of all kinds are subject to nest depredation by many different predators. Numerous species of reptiles, mammals, and yes, other birds, just love a healthy breakfast of fresh eggs or baby birds, so songbirds have evolved strategies to dodge nest-raiding predators. For example, concealment of the nest and camouflaging plumage in female birds is ubiquitous throughout the avian world. Likewise, songbirds themselves are heavily preyed upon by certain raptors—accipiters, such as Sharp-shinned Hawks and Cooper's Hawks—are bird-killing specialists. But many species of owls also routinely hunt birds, both during daylight and crepuscular hunting forays, when other birds need to be especially wary, but also at night if owls find birds that are roosted.

Perhaps the ingrained urge to protect both themselves and their nests is what, evolutionarily speaking, drives small songbirds to routinely harass any owl they find roosting during the day. Boisterous species such as chickadees,

A Northern Mockingbird expresses its dislike of a Great Horned Owl.

nuthatches, titmice, gnatcatchers, and jays are especially adept at sounding the alarm when they discover a roosting owl, and when one or two small birds begin excited scolding, soon other birds join the fray. Owl mobbing can become a frenetic multispecies cacophony.

Thanks to the excited scolding calls of Black-capped Chickadees, I once discovered a roosting Western Screech-Owl wedged into a narrow nook 10 feet up in an aspen tree about 50 feet from a creek I was fishing just after sunup. I decided to sit quietly at a respectful distance and watch the drama unfold. I couldn't determine how long the owl had been under assault from two chickadees (and, as it turns out, a Nashville Warbler) because I'd worked my way around a bend on the gurgling stream before coming into earshot of the fracas. But as I sat there very still, two more chickadees arrived, followed by two Red-breasted Nuthatches, and then a Yellow-rumped Warbler. The Nashville Warbler had by then disappeared, and the Yellow-rumped Warbler appeared only briefly and covertly.

The chickadees were boldest; the warblers kept their distance. After perhaps five minutes, a Ruby-crowned Kinglet boisterously joined the fray, and then came two very loud chattering Cassin's Vireos. The mobbing reached its crescendo and then steadily tapered off as birds slowly secreted away, leaving only two chickadees to continue the onslaught. I sat for a while longer because, while I'd seen other owl-mobbing incidents, I'd never considered just how long such an affront to the owl's privacy might last. This one went on for about 10 minutes, but only at full multispecies strength for a minute or two. Soon an American Robin arrived on the scene, very agitated. Then things quieted down. So, I wondered: would the owl be subjected to such mobs routinely throughout the day, or had it borne the brunt of avian ire for the time being?

Not surprisingly, scientists have studied mobbing behavior and discovered some answers in experiments conducted by two researchers from Old Dominion University (Virginia) in the 1980s. They used three different methods to elicit mobbing behavior from songbirds: an Eastern Screech-Owl mount from a taxidermist in conjunction with recorded Eastern Screech-Owl calls, the screech-owl mount without associated calls, and screech-owl calls without the mount. The experiments took place in situ—in the woods within screech-owl habitat, and the team undertook a total of 508 trials. When they combined the mounted owl (placed in a tree where a typical screech-owl might sit out the day) with the recorded owl calls, mobbing occurred in more than half of the 170 trials; with the calls only (no mounted owl), mobbing occurred in 35 percent of 169 trials. Across all the trials in these two scenarios, a total of about five dozen different species of birds responded. But when the owl decoy was left in situ with no accompanying taped owls calls (169 trials), potential mobbers rarely found it and never mobbed it.

Their results suggest that songbirds innately recognize predator calls (in this case calls of the Eastern Screech-Owl). Moreover—and intriguingly—the experimenters concluded, "Finally, we can speculate on the significance of these results for the behavior of owls. Our results suggest that owls perched silently at a day roost are detected and harassed rarely by mobbing birds. A calling owl, however, will almost certainly attract potential mobbers (approximately 85 percent of trials in this study). Because owls that are detected by birds are usually mobbed relentlessly, and presumably suffer reduced hunting success or increased risk of predation, mobbing behavior may be an important force in selecting for cryptic diurnal behavior (e.g., silence, well-concealed day roosts, etc.) in some owls. Conversely, mobbing may encourage owls to restrict conspicuous behavior, such as calling, to nocturnal hours. If these suggestions are correct, mobbing may

Probably the first illustrated instructions for using live owls as decoys to capture songbirds appeared in the Italian-language *Uccelliera*, by Giovanni Pietro Olina, published in 1622. This art piece shows numerous twigs, covered with birdlime, deployed to capture the unwitting songbirds mobbing the owl.

reinforce nocturnal behavior in predators such as screech-owls and thereby reduce the probability of future encounters between a mobber and its potential predator."

But . . . Be Careful Out There!

Additional research on mobbing behavior, published in 2022, revealed that mobbing behavior aimed at Northern Pygmy-Owls peaked at times when risk to individual mobbers was lowest, namely during late summer and fall when songbird populations are bolstered by lots of juveniles—an abundance of potential mobbers reduces the risk to each individual mobber. Researchers at Oregon State University (OSU) conducted trials within Northern Pygmy-Owl habitat, using recorded owl calls to elicit mobbing responses by songbirds. These tests, conducted across all seasons and at varying elevations, revealed that mobbing behavior was uncommon, but more frequent during late summer and fall.

Northern Pygmy-Owls are bird eaters that hunt by day, and any small bird up to about the size of the tiny owl itself is potential prey. So, mobbing a pygmy-owl carries risk to the mobbers—they need to be mighty careful that they don't end up on the owl's dinner plate. During much of the year, pygmy-owls prey heavily on birds, but come winter and spring, when food is scarce in the absence of fledglings and of many migratory songbirds that don't return until April and May, the owls expand their diet to include lots of small mammals. The relative risk to songbirds decreases. During summer's abundance (including abundant young-of-the-year songbirds), the owls return to a bird-dominated diet, so all little birds better beware.

Moreover, mobbing requires a lot of energy on the part of the songbirds—it's a behavior that tends to be frenetic. Burning up calories during summer and fall, when food is abundant is one thing; but the scarcity of winter and early springs puts most species on the razor's edge of the energy equation: burn more calories than you consume, and you risk starvation. The OSU scientists hypothesize that mobbing behavior is most common when Northern Pygmy-Owls pose the greatest threat to songbirds—late summer and fall—and when energy expenditure on mobbing behavior is the least perilous.

So, mobbing can certainly be risky, but not nearly so potentially perilous as it was in centuries past. In olden times (and still today in a few places), songbirds were collected for food, research, sale, and feathers by luring them to branches covered in *birdlime*, a catch-all word for any number of different sticky substances—the equivalent of modern-day glue traps and every bit as inhumane. But first the bird trappers had to attract the birds they sought to capture, and a favorite method for doing so was to use a live owl, tethered to a perch, to take advantage of owl-mobbing behavior. In some places, the owls were even trained to perform this duty, as reported by Italian ornithologist Tommaso Salvadori in Henry Eeles Dresser's *A History of the Birds of Europe* (published between 1871–1896). The Little Owl (*Athene noctua*), Salvadori explained, was "very much used in Central Italy by sportsmen as a decoy for Larks, which it attracts within range of the gun, or for small songbirds, which congregate in numbers to witness its grotesque attitudes and to hear its mimicry; thus they are easily snared on limed twigs."

Owlish Scarecrows Don't Scare Crows

Have you ever seen those life-size plastic owl decoys mounted on rooftops and similar highly visible locations? They are often used to scare away pigeons. In 1986, these plastic scare-pigeons even made the news in the illustrious *New York Times*. Residents and businesses in the pigeon-poop-plagued Big Apple had turned to plastic owls to alleviate the problem. As it had for decades in many venues, the ploy worked—for a while. But after a time, the city-slicker pigeons realized these owls never moved, not so much as a ruffling of a feather. And at that point

the owl forgeries became pigeon perches, and of course, pooped-on perches. The game of one-upmanship began, as people started devising ways to animate the plastic owls.

Luckily, they never resorted to the older practice, conceived in the days before plastics (and before raptors were protected), of using live or stuffed owls as decoys to scare away a variety of so-called vermin—or, in an ironic twist, to actually *attract* vermin to within range of a shotgun. For example, just when a Kentucky farmer was at wit's end over his inability to prevent massive flocks of crows from descending upon his crops and leaving them in ruins, he stumbled upon a potential solution. "Year by year was he molested by these depredators, which would tear up his corn by the acre, pick out the eyes of his lambs, fly off with his young chickens, and annoy him in every conceivable way," reported the *Farmer's Register* in 1841.

No matter how stealthily he tried, he could never get within gun range of the wary and intelligent crows, but one day he happened upon a chaotic scene: numerous crows had discovered a large owl and were pestering it raucously. "Our farmer was too interested a spectator of the combat not to reflect much upon its character and result, and all at once it occurred to him, that, if by any means he could get possession of an owl, he could make him decoy the crows within a reasonable distance."

By means illegal today, the farmer did in fact capture an owl and immediately set about putting the beleaguered raptor to work for him, tethering it to a prominent perch, then concealing himself nearby. "Scarcely were these things completed, before a distant and well known caw broke upon his ear, and anon the air was darkened with a flight of crows."

The decoy owl worked beyond the farmer's wildest expectations, and soon the pesky crows were falling to his gun by the brace, their numbers unable to resist the innate desire to harass the owl. Moreover, the owl seemed to revel in his job, "for, at each dreary pause, while the farmer was reloading—the sagacious captive would ruffle his feathers and snap his bills together, and manifest to his enemies the most aggravating and insulting behavior. This would exasperate them beyond bounds, and at him they would come again—bang would go the gun, and at every crack the owl fairly chuckled with delight, giving one of those knowing winks."

Finally, the farmer desisted, too weary to go on, and he took the owl home and gave it a feast of slain crows. He repeated the performance daily until nearly every crow on his plantation was killed. By then, word of the feat had spread, and the owl's fame had grown to legend. The farmer lent his owl to neighbors so they too could rid their plantations of marauding crows. "As may be supposed, good

The Northern Pygmy-Owl has ocelli—false eyes—on the back of its head, and scientists continue to study their functions.

Eyespots are common in the animal world. Many moths and butterflies have them, including this aptly named Tawny Owl Butterfly.

care was taken of the owl, and for two seasons he was the greatest benefactor of the neighborhood, and had been the death of as many of his foes as Genghis Khan or Napoleon."

Eyes in the Back of Your Head?

Many different animals have false eyes, sometimes called *ocelli* or eyespots—think about all the butterflies and moths you've seen with wings adorned with round, often colorful, haloed spots. The reasons for eyespots aren't entirely understood and serve different functions for different species, but among birds of prey, pygmy-owls sport the most well-defined false eyes—patches of black feathers, haloed in white, on their napes. For a long time, scientists reasoned that eyespots might help animals ward off predators by making the potential victim look bigger or more alert, and because a pygmy-owl's false eyes are on the back of its head—its blind spot—this idea made sense.

But theories are meant to be tested, and in the late 1990s Denver Holt—founder of the Owl Research Institute—designed a study to determine the role of a pygmy-owl's false eyes. University of Montana grad student Caroline Deppe took the lead role in the study, using wooden pygmy-owl decoys to elicit mobbing behavior by other birds. Northern Pygmy-Owls and other pygmy-owl species prey heavily on songbirds, so they are often the targets of such mobbing behavior. Turns out that those false eyes on the owl's nape might serve the purpose of fooling mobbing songbirds into exerting their efforts on the front side of the owl, where it can keep an eye on them, and also might reduce the duration of mobbing behavior and reduce the number of mobbers.

Deppe reported, "When mobbers made close passes around the Northern Pygmy-Owl models, they avoided the eyespots and passed in front of the true eyes. When mobbers made close passes around the non-eyespot model, they performed the behavior equally to the front and back of the model. From the Northern Pygmy-Owl's perspective, eyespots appear to direct mobbers to a location where the owl can see them."

Mobbing birds sometimes strike and injure pygmy-owls, so seeing your enemy is certainly advantageous, but so too is being able to see potential prey—the pygmy-owl's eyespots might serve both purposes. A few years after Deppe published her findings, researchers from Spain and Canada furthered the study of pygmy-owl eyespots and determined that "species displaying ocelli [eye spots] in the nape tend to have a high proportion of small birds in their diets and live in open habitats, whereas the opposite is true for species without ocelli. Pygmy-owls with ocelli are also considerably smaller and, collectively, these findings are most consistent with false faces being a conspicuous visual signal to deceive mobbing birds so they can be preyed upon."

Worldwide there are more than two dozen species of pygmy-owl, and many if not most are diurnal hunters that prey heavily on songbirds. So perhaps those eyespots passively serve to protect the tiny owls from becoming victims of their mobbers, while actively helping them lure prey into range or even evaluate potential hunting grounds.

Branchers on the Family Tree

For most any baby bird, that first venture beyond the nest is fraught with peril—wings are not fully developed and have never been tested; legs and talons are not quite ready for everyday duties. But the time has come to leave the nest, so they carry forth, sometimes bravely, sometimes with trepidation, and the avian world is full of fly-the-coop strategies: tiny fuzzball Wood Duck chicks perching upon the lip of a tree-cavity nest hole, then jumping to earth or water—which might be 50 feet below; or, more perilous still, the chicks of cliff-nesting Barnacle Geese plunging hundreds of feet to their waiting parents, many of the babies perishing in the effort.

Many species of baby owls stay close to home upon leaving the nest, perching upon a nearby tree limb. This is called the *branching* stage, and the owlets are called *branchers*.

These juvenile African Barred Owlets are old enough to leave the nest but not yet ready to fend for themselves. With most woodland owl species, fledglings of this age go from the nest to a nearby branch as they start learning to use their feet and wings—hence the terms *branchers* and *branching owls*.

They are not yet able to fly, but their feet are at or near full size and already very strong. They can grip branches and even walk and climb as needed, but the little owlets are still dependent upon their parents for food. If you are lucky enough to come upon two, or three, or four fuzzy, wide-eyed owlets lined up on a horizontal branch, keep your distance and don't linger. The owlets may seem mesmerized by your presence, but they cannot readily escape, and if disturbed they may become vulnerable to predators. One or both parent birds are likely nearby keeping an eye on the kids and on you. Likewise, if you find a baby owl on the ground or on a low branch, it's usually best to leave it be unless its life is in jeopardy from traffic or domestic animals such as dogs or cats; many branchers end up at wildlife rehab facilities because well-meaning Samaritans think the little owlet is in need of rescue when in fact the parents, the nest, and the branching limb are all probably nearby.

When in Rome (or Rather, Iceland)

One of the owl heavyweights, the Snowy Owl is an arctic icon, a circumpolar species that nests on tundra during the long days of the far north summer. These beautiful white birds are tough customers, able to thrive in a brutal environment: the arctic tundra is the coldest of all biomes on Earth. The average annual temperature in Barrow, Alaska, for example, is 10.4 degrees Fahrenheit; those balmy days of summer, the time of the midnight sun, reach into the 40s, but during the long winters, temperatures in Barrow rarely climb above zero.

Snowy Owl with the remains of a large bird

Throughout their worldwide range, Snowy Owls feed primarily on small rodents, especially lemmings. Iceland has no lemmings but does have a small population of the beautiful white owls of the far north. So, what's an owl to do in the absence of lemmings? Well, in Iceland at least, other birds better beware when a Snowy Owl is on the prowl. Studying the summer diet of Snowy Owls in Iceland, researchers discovered that birds are the primary prey—especially Rock Ptarmigan (a species of grouse that turns white in winter) and shorebirds (e.g., sandpipers), but also waterfowl and songbirds.

Luck of the Lemming

On the other hand, over most of its circumpolar breeding range, Snowy Owls feast on lemmings—to the tune of three to five of the fuzzy little rodents daily. Lemmings comprise some 19 species worldwide, most of which live in the arctic—the haunt of the Snowy Owl. Lemmings are the most important factor in Snowy Owl brood success, and populations of these cute little rodents vary annually, and often dramatically. In fact, the lemming boom/bust cycle can yield bumper crops of baby owls one year and complete local or regional reproductive failure the next, as the Owl Research Institute discovered in its long-term Alaska Snowy Owl research project. In 1994, surveyors didn't find a single owl nest within the 100-square-mile

Lemmings are a critical food source for Snowy Owls.

study area; the next year, they logged 54 nests. The difference was lemmings—lots and lots of lemmings in 1995, after a lemming crash the previous year. Other factors impact nesting success and brood survival for Snowy Owl parents, but lemmings are the cog.

How to Hire . . . an Exterminator?

Reportedly, Eastern Screech-Owls (and Elf Owls) in the Desert Southwest will capture Texas Blind Snakes (*Rena dulcis*), carry them to the nest cavity, and deposit them there alive, where the snakes can serve as unwitting exterminators, eating parasitic arthropods that are harmful to eggs and baby owls. Blind snakes, also called thread snakes, are small, wriggly, primarily subterranean snakes that look like shiny earthworms. They are rarely more than a few inches long but can reach about 10 inches in length; their eyes are little more than two black spots under the scales on the head.

In the 1980s, Baylor University researchers who found these snakes alive in several owl nests posited an adaptive behavior by the owls. However, they also reported that many such snakes are damaged from capture, perhaps indicating a case of beneficial coincidence rather than programmed behavior: maybe the owls capture the snakes as food for their owlets, but fail to kill the snakes, which then bury themselves in the nest substrate and subsist off invertebrates in the nest. Normally, owls kill their prey immediately, and as the study authors pointed out, "reptilian prey often dangles from the bill of adult screech-owls upon delivery to the nest, but the 4 live blind snakes were coiled about the bills of the owls that carried them."

This defense technique may have saved those four snakes from being killed, so rather than an adaptive behavior by adult screech-owls, perhaps the snakes were lucky to be alive and because they lived, baby owls benefitted. Live blind snakes living in the substrate of the owl nest cavities fed on insect larvae associated with owl feces, owl pellets, and prey remains. Thanks to these slithery housekeepers, owls with "live-in blind snakes," reported the researchers, grew faster and experienced lower mortality than same-season screech-owl broods lacking live blind snakes. As biologist Enrique H. Bucher pointed out in a 1988 article in *Parasitology Today*, "the apparent benefit to the owls of bringing an efficient predator of insects to their nest is attractive, and certainly merits further study, although it is difficult to understand why such supposedly advantageous behavior is so restricted in terms of the number of species involved and geographical range."

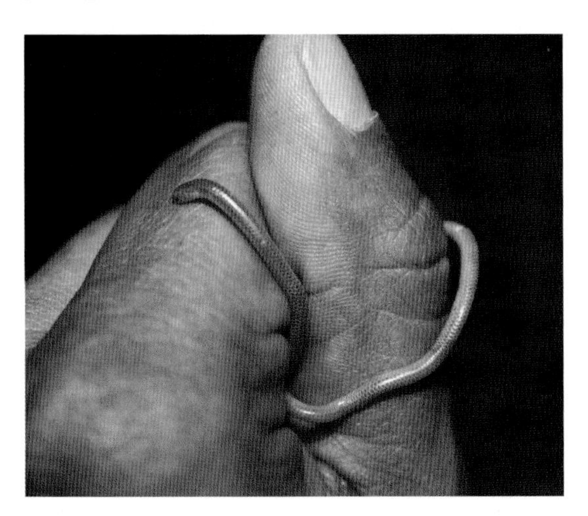

Blind Snakes superficially resemble earthworms, and like earthworms spend most of their lives underground.

Eastern Screech-Owl

NOT SO WISE AFTER ALL?

The ferocious, though decidedly cute little bantamweight we call the Boreal Owl is a denizen of high latitudes worldwide. In North America, the Boreal Owl was named Richardson's Owl by French ornithologist Charles Lucien Bonaparte to honor arctic explorer and naturalist Sir John Richardson, but the handsome little owl was already known to indigenous people for its stunning lack of judgement.

The Innu people (named Montagnais by the French) who lived in what is now Labrador and Quebec called this owl *pillip-pile-tshish*, which means "water-dripping bird," referring to one of the owl's calls: in 1882, ornithologist Clinton Hart Merriam reported, "This owl has a low liquid note that resembles the sound produced by water slowly dropping from a height."

But the water-dripping bird was once the largest owl in the world, with a thunderous voice, says the Innu legend. So loud and proud was the massive owl that one day it perched near a large waterfall and not only tried to imitate the sound of the plunging water but also attempted to "drown out the roaring of the torrent with its own voice," according to Merriam's retelling. Such pomposity, however, offended the Great Spirit, who transformed the giant owl into the pint-size bird we know today, "causing its voice to resemble slowly dripping water instead of the mighty roar of a cataract."

Boreal Owl, from *Svenska Fåglar* (Swedish Birds), 1838, by Magnus Von Wright et al.

Postcards from Mexico

What's an owl to do if it doesn't like cold winters? Simple—fly south. Several North American owl species are migratory, bailing out from the cold north before Old Man Winter casts his icy spell. Migration distances and winter ranges vary by species. For the Long-eared Owl and Short-eared Owl, just vacationing in the southern parts of their overall range is enough, and most of them overwinter in the Lower 48. Burrowing Owls breed as far north as Central Canada, but these northern nesters split town for winter quarters in the American Southwest and

Mexico. The aptly named Elf Owl, likewise, expands its range northward to Arizona, New Mexico, and Texas during nesting season, but soon retreats back to Mexico.

But perhaps the most impressive migratory owl is the Flammulated Owl, a bird of montane forests. This insect-eating, loud-mouthed mighty-mite spends the summers in the mountains of the West, from southern British Columbia to northern Mexico, with populations inhabiting most all the mountain ranges of the American West, from Colorado to California and New Mexico to Washington. Its surprisingly deep and resonant hoots sound like they should be coming from a much larger owl: the Flam, as these little nocturnal forest dwellers are called by their fans, stands a mere 6 inches tall. And they're the ultimate snowbirds, shunning those high-elevation winters when insect prey becomes terribly scarce. Flams spend the cold months basking in the Mexican sun, with some even venturing farther, to Central America. Moreover, like most birds, these tiny owls migrate at night: specimens captured in mist nets by researchers are invariably nocturnal travelers.

The Wonders of Wandering

While a few of North America's owls are migratory, several others can be both nomadic and irruptive. The term *irruptive* refers to a species that occasionally expands its normal range during winter, primarily southward. Often attributed to food shortages, irruption is a bit more complicated than that. The boreal forests of the far north periodically produce bumper crops of pinecones and seeds (an event called "masting"), giving songbirds such as Pine Siskins, Red-breasted Nuthatches, crossbills, and Pine Grosbeaks an abundant food supply, which leads to high fecundity. With way more birds occupying the same habitat, food supplies run low, and birds begin to fan out southward in large numbers. That's why some years you might have dozens of Pine Siskins at your birdfeeders during winter and other years none at all.

Similarly, owl irruptions seem to happen after high-productivity nesting seasons, which occur when prey is abundant. Among owls, Snowy Owls—the gorgeous white owls of the arctic—are the best-known "irruptors." Every few years, quite a few of them wander south to southern Canada and the northern tier of states; and every once in a long while, a Snowy Owl makes headlines much farther south during mega-irruptions. A high proportion of irruptive Snowy Owls are juveniles, but as Project SNOWstorm researchers discovered, "snowy owls in major irruption years tend to be fatter and heavier than those in non-flight years. Only occasionally do food shortages appear to prompt southerly movements of snowy owls."

Other northern owls, particularly Northern Hawk Owls, Great Gray Owls, Short-eared Owls, and Long-eared Owls, can also irrupt, but typically these events include more adult owls and may thus be driven by food shortages rather than high brood success; however, these two phenomena are closely linked and causative—lots of food often results in lots of baby birds.

In 2012, which was a significant irruption year for Snowy Owls, this individual spent much of the winter in southeast Oregon.

Whether it's an owl or a songbird species that suddenly expands its wintering range, irruption remains somewhat enigmatic, but at least for Snowy Owls, these mass movements seem to be driven by high nest success at least somewhere in the breeding range. High fecundity years, driven by abundant food, lead to lots of young owls, which in turn creates lots of competition for food. In other words, suddenly lots of Snowy Owls occupy an area that may not proffer enough winter food to support such numbers, but even that supposition remains under scientific scrutiny. Despite decades of studying a population of Snowy Owls in Alaska, the Owl Research Institute says, "There is a long-standing myth that irruptions are driven by starving owls. In reality, one of the only things we know with certainty is that irruptions are indicative of a strong breeding season somewhere in the Arctic."

Something Fishy Here!

All owls are carnivores, and most are generalists, willing to kill and eat a wide range of prey—insects and arthropods, reptiles and amphibians, mammals, and even other birds.

DELIGHT IN EVERY BITE

Although the various fish-owls and fishing-owls specialize in aquatic prey, many other species happily indulge in a little waterside dining when the opportunity arises. In North America, for example, Barred Owls wade through shallow water to snack on crayfish, frogs, and even the occasional fish. In fact, just about any living creature of appropriate size is in danger from owls of one kind or another, and sometimes—during the harsh winters when food is scarce—owls will even eat carrion.

Most owls feed heavily on small mammals. Some of our most familiar species—the Great Horned Owl, Barn Owl, Barred Owl, Great Gray Owl, Snowy Owl, Long-eared Owl, Short-eared Owl, and even the comparatively diminutive screech-owls—reign terror upon voles, mice, and other little rodents. And the bigger the owl, the bigger the mammals they can capture. Great Horned Owls, which occupy a greater diversity of habitats than any other owl in North America, prey on small rodents but also on rabbits, hares, opossums, squirrels, and even skunks. They eat just about anything, and like most owls, their diet varies with location and time of year; while small mammals are mainstays, in some places Great Horned Owls feed heavily on reptiles, amphibians, insects and other invertebrates, and especially birds (including other owls, as well as hawks, waterfowl, wading birds, songbirds, and game birds).

Barred Owls are adept at catching crayfish from shallow water.

One family of Great Horned Owls living on 379-acre Protection Island in western Washington State dined only on birds and almost exclusively on Rhinoceros Auklets (a small seabird that nests on the island) during two consecutive summers when scientists collected and analyzed pellets. None of the 129 pellets contained any mammal bones. One witness interviewed by the researchers reported seeing an owl capture an auklet from the entrance to its nesting burrow on the island, and another time watched an owl snatch an auklet from the water near an adjacent island. Heading out to saltwater to capture food for their chicks, Rhinoceros Auklets depart from (and return to) their nest burrows under cover of darkness, making them vulnerable to Great Horned Owls. Outside of the summer auklet nesting season (when the little seabirds live far offshore), these same owls were known to kill and eat invasive peafowl that inhabited the island—and peafowl (aka peacocks) average two or three times heavier than Great Horned Owls.

On the other end of the prey scale, Great Horned Owls and other large owls enjoy appetizers in the form of bugs and other little creepy crawlies. But many small owls, including the 6-inch-tall Flammulated Owl of the Western mountains, are insectivorous, meaning they primarily eat insects and other small invertebrates. Endearingly called Flams, these tiny owls of evergreen forests pluck insects off the ground, glean them from foliage, and even snatch them out of midair like a nocturnal flycatcher.

Great Horned Owls prey on many creatures, in this case a cottontail rabbit.

Most owls eat insects and other invertebrates, and some small species dine almost exclusively on them. This Tropical Screech-Owl has captured a fearsome looking insect in Colombia.

Buffy Fish-Owl (*Ketupa ketupu*)

But a few owls specialize in fishing. The four species of Asian fish-owls and the three species of African fishing-owls all prey heavily on fish, along with other aquatic creatures. Primarily they hunt by wading through shallow water or perching on limbs or rocks above and near the water, waiting and watching for a chance to snatch fish, frogs, and other animals. Consequently, the fish- and fishing-owls have physiologically diverged somewhat from typical owls: fish- and fishing-owls have poorly defined facial disks, suggesting that hunting by sound is not as important as for other owl species; likewise, they lack the comblike edges to their flight feathers, indicating that they don't rely on silent flight—which makes sense when the typical prey is found beneath the surface of the water. Fish- and fishing-owls also have toes adapted to grasp slippery prey. The Blakiston's Fish Owl (*Ketupa blakistoni*) of Japan, China, and eastern Russia is the world's largest owl, and extensive research into this enigmatic bird led to Jonathan C. Slaght's award-winning book *Owls of the Eastern Ice* (2020). The other fish-chowing owls are the Brown Fish-Owl (*K. zeylonensis*), Tawny Fish-Owl (*K. flavipes*), Buffy Fish-Owl (*K. ketupu*), Pel's Fishing-Owl (*Scotopelia peli*), Rufous Fishing-Owl (*S. ussheri*), and Vermiculated Fishing-Owl (*S. bouvieri*).

All in a Day's Work

Despite their reputation for haunting the night, many owls are frequently more active just before dawn and just after dusk than they are in the wee hours, but some owl species are sun lovers by comparison. Several species of pygmy-owls—including the Northern Pygmy-Owl found in the United States—are primarily daylight hunters. As such, they lack the asymmetrical ears that allow nocturnal owls to pinpoint prey by sound, and they don't have the specially modified feathers that help many owls fly soundlessly. Nonetheless, these pint-size predators have extraordinary eyesight and agile flight, and while they typically wait and watch from a favorite perch, they are amply capable of lightning-fast attacks even on other birds.

Their far-north cousin, the imposing Northern Hawk Owl, is likewise prone to hunting in broad daylight and in both appearance and behavior resembles a falcon, with its swift,

The Short-eared Owl frequently hunts during daylight, coursing over prairies, fields, and marshlands in search of rodents.

direct flight and long, pointed wings. The Northern Hawk Owl occasionally preys on birds (a falcon specialty) as large as grouse, which outweigh the sleek owl, to supplement its diet of small mammals. Of course, owls of the far north, most notably the iconic Snowy Owl, have little choice but to be adept at both diurnal and nocturnal hunting given the scant daylight of winter and the fleeting nights of summer.

Ranging across much of the continent, Short-eared Owls frequently hunt by day, especially during mornings and evenings, and are especially active in daylight when they have hungry chicks to feed in springtime. In fact, many owl species will extend hunting hours into the morning or start a little earlier in the evening when nestlings demand a constant supply of food.

The Barred Owl has expanded its range westward across North America to reach the Pacific Northwest, where it poses a serious threat to the survival of the closely related Northern Spotted Owl.

The Beast from the East

Throughout its range in the Pacific Northwest, the Northern Spotted Owl battles for its very existence. This cryptic owl's crucial habitat—expansive tracts of old-growth montane conifer forest—has been extensively fractured by commercial timber harvest, especially from clear-cut logging. And just when the Spotted Owl's plight was leading to at least a modicum of better timber-harvest practices and laws to protect some of its habitat, a new threat arrived in the form of the closely related Barred Owl.

Barred Owls are historically native to eastern North America, but beginning early in the 1900s, the species began marching its way westward, greatly expanding its range. Researchers postulate that these forest-dwelling owls had been barred, so to speak, from the Great Plains, owing to routine burning by Native Americans prior to the arrival of Europeans. Native people set blazes to drive prey animals, particularly bison, and perhaps to manage grazing-animal habitat. But with Native cultures decimated after their lands were permeated by Euro-Americans, thousands of years of prescribed burning came to an end. Trees began to grow, especially along waterways, creating a trail of forests for Barred Owls. By the 1950s, they had reached British Columbia, and then they poured southward, finding ample wooded habitat throughout the verdant Northwest all the way to California.

Now permanent residents of the West, Barred Owls are remarkably adaptable, able to thrive in suburban areas as easily as in wilderness. They are ferocious generalist predators that will hunt and kill a wide range of prey; like many owls, they eat lots of small rodents but will opportunely take many other animals—birds, reptiles, amphibians, shrews, insects, and even crayfish. Moreover, Barred Owls sometimes exhibit their ferociousness by attacking humans who venture too near a nest or fledglings—as was the case at Salem, Oregon's 90-acre Bush's Pasture Park in 2015 when the local pair of Barred Owls attacked no fewer than four people; and in 2022, a Washington woman was attacked by a Barred Owl twice in one November week in her own driveway. Thereafter she began carrying an umbrella to shield her head from the silent assassin.

Barred Owls are subject to intensive study in their newly pioneered western real estate because the entire population of Northern Spotted Owls is shrinking rapidly in the face of these invaders (invasive because we inadvertently gave them a pathway). Already imperiled by significant loss of habitat and habitat fragmentation, Northern Spotted Owls are extremely vulnerable to dramatic ecosystem changes such as encroachment from a competing species. Studies demonstrate that Spotted Owl declines are most significant in areas where Barred Owls have moved in and where they've been the longest. So now it's a race against time to find out if scientists and managers can conceive and implement plans that will save the Northern Spotted Owl before it's too late.

Forefront in the U.S. Fish and Wildlife Service's Barred Owl management strategy is the removal of Barred Owls from Spotted Owl territories. This method proved successful in halting declines of Spotted Owls at five study areas in the Pacific Northwest. Ultimately this multiagency study, which began in 2002, demonstrated a positive correlation between Barred Owl removal and reduced Spotted Owl population loss. A 2021 study abstract explains, "Removal of barred owls had a strong, positive effect on survival of sympatric spotted owls and a weaker but positive effect on spotted owl dispersal and recruitment. After removals, the estimated mean annual rate of population change for spotted owls stabilized in areas with removals (0.2 percent decline per year), but continued to decline sharply in areas without removals (12.1 percent decline per year). The results demonstrated that the most substantial changes in population dynamics of northern spotted owls over the past two decades were associated with the invasion, population expansion, and subsequent removal of barred owls. Our study provides experimental evidence of the demographic consequences of competitive release, where a threatened avian predator was freed from restrictions imposed on its population dynamics with the removal of a competitively dominant invasive species."

Unfortunately, *removal* is a euphemism: the removal method for Barred Owls is a 12-gauge shotgun. Beginning in 2015, researchers shot nearly 2500 Barred Owls, but lethal removal seems to be the only option for saving native Spotted Owls in the Northwest. In 2019, Bob Sallinger, who was longtime conservation director for Portland Audubon, told news media, "A decision not to kill the barred owl is a decision to let the spotted owl go extinct."

The plan is controversial, of course, because of the inherent ethical dilemma: killing a non-native species to save a native species. The Barred Owl has no moral culpability—its range expansion was caused by humans—but it bears the lethal brunt. With luck, killing off the new arrivals in key tracts of Spotted Owl range will save this alluring bird from obliteration. Only a few thousand Spotted Owls remain on federal lands in the Pacific Northwest (and most of their remaining habitat is on federal lands), so saving this iconic bird is a race against time—and ultimately, we must hope the Barred Owl's sacrifice is not in vain.

The Beast of the East

The world's largest owl is the Blakiston's Fish-Owl of northern Japan, China, and the Russian Far East. Weighing 6.5 to 10 pounds, with females averaging about 25 percent larger than males, these heavyweights of the owl world nest in large tree hollows, such as broken-topped old-growth trees and snags in riparian areas. They need open water for hunting, and in the far north, such places are typically stream reaches with sufficient current to prevent ice-over or with springs providing above-freezing water. Blakiston's Fish-Owls feed mostly on fish, including pike, salmonids (salmon, trout, and char), sculpins, and lamprey, but also take mammals, crustaceans, amphibians, and occasionally birds. Though the fish-owl moniker is apt given their diet, Blakiston's Fish-Owls seem most closely akin to the Eurasian Eagle-Owl (the heaviest of which can slightly outweigh the largest fish-owl), but its genetic lineage and thus its taxonomy remains under study. One thing we know for certain about this enigmatic giant is that it is imperiled; the entire global population may consist of only a thousand or so individuals. They are classified as an endangered species by the International Union for the Conservation of Nature (IUCN), and face myriad threats, including widespread habitat loss. Luckily, the Blakiston's Fish-Owl received significant fanfare from one of its ardent champions, researcher Jonathan C. Slaght, when he authored *Owls of the Eastern Ice* (2020). The book chronicles how Slaght became mesmerized by the plight of this little-known species and devoted his professional life to its well-being, beginning with excruciatingly challenging research projects in remote eastern

The endangered Blakiston's Fish-Owl is the world's largest owl.

Russia. Slaght also cofounded the Blakiston's Fish Owl Project (FishOwls.com), the leading repository for information about this remarkable bird.

And Speaking of The East . . . An Owl Saves the Day!

The subject of innumerable fables and legends, Genghis Khan (1162–1227), the famous Mongol king, is perhaps solely responsible for the reverence with which owls have long been treated by the descendants of his once-mighty Asian empire. According to legend, an owl once saved the life of the great universal ruler, who at the peak of his omnipotence reigned over the largest contiguous land empire the world has ever known. Through force and persuasion, he brought all the Mongol tribes under his rule and conquered vast territories throughout China and Central Asia. Not surprisingly, Khan's domination earned him no modest supply of enemies.

About a century after the Great Khan's death, a popular book circulating in Europe—*The Travels of Sir John Mandeville* (actual authorship unknown)—told a story that persists to this day. Genghis Khan had ridden with a few of his men to survey lands recently conquered, but the small contingent came face to face with a large enemy force. The Great Khan and his men had no choice but to retreat, and the chase was on; the enemies managed to kill Genghis Khan's horse from beneath him. His men thought he too had been killed and continued their hasty retreat, while their leader made his way into a dense copse of trees and shrubs and secreted himself away, hoping to avoid detection.

The enemy force continued in pursuit of the Mongols, but upon searching the woodlot where Genghis Khan was hiding, they saw an owl perched in a tree and reasoned that one would be foolish to hide beneath an owl, an omen of foreboding. So they moved on in their quest to hunt down the invaders,

Legend says an owl once saved Genghis Khan from being captured and perhaps killed by his enemies.

"and thus," says the 14th-century account, "escaped the emperor from death. And then he went privily all by night, till he came to his folk that were full glad of his coming, and made great thankings to God Immortal, and to that bird by whom their lord was saved. And therefore principally above all fowls of world they worship the owl; and when they have any of their feathers, they keep them full preciously instead of relics, and bear them upon their heads with great reverence; and they hold themselves blessed and safe from all perils while that they have them upon them, and therefore they bear their feathers upon their heads."

The earliest surviving version of *The Travels of Sir John Mandeville* is in French, but the work was translated into many languages, so the details of this story vary somewhat, but all accounts agree that the owl was then venerated by the Mongols. Finnish owl expert and author Heimo Mikkola says, "Genghis Khan then adopted the owl as a good luck charm—from then on, he and his followers wore owl feathers and charms both to protect themselves from danger and pay tribute to their special saviour," and explains that "to this day, people of the Mongol Steppe venerate owls. . . . In Mongolia, as in many parts of Asia, the owl is considered a protector and a divine ancestor, who helps to ward off evil spirits, famine and pestilence."

By Any Other Name

Many Native American peoples revered owls to the extent that these birds were strongly associated with shamans—the spiritual leaders and healers. *Shaman* is a word often used in literature to describe a host of similar aboriginal roles, generally correlating to medicine man or medicine woman, doctor, healer, spiritual leader, sorcerer, mystic, and others. Throughout the Americas, there were hundreds if not thousands of Native-language terms for shaman, not surprising given the hundreds of languages spoken in the precontact New World. Shamans often carried stuffed owls, owl feathers, and owl amulets. They trusted the owl as a messenger or advisor, and by evoking the owl they too hoped to be able to see in the proverbial dark, to ascertain wisdom unavailable to others.

For the original Americans who venerated owls, these enigmatic birds frequently served as escorts to the spirit world and as harbingers or foretellers of impending misfortune, for which the owl itself is held blameless. Such cultures often believed owls also served as watchful guardians. Eighteenth-century missionary John Heckewelder, having spent considerable time among the Lenape people of the Northeast, reported that on his numerous excursions in the woodlands with them, if an owl called, a tribesman would immediately rise and throw some tobacco in the fire. Upon asking about this tradition, Heckewelder was told that the offering was because the owl had come to "guard over them by night, for they gave them warning, whenever an enemy approached, or was about to surround them, especially when at war."

Although traditions vary widely given the astonishing number of cultures that arose in the New World over thousands of years, Native Americans often viewed owls as less frightening and ominous than did Europeans, and they were well versed in all the owl species. Native Americans were intimately connected to nature, far more so than agrarian and urban Europeans had been, for thousands of years. As one linguist, studying indigenous names for owls in Mexico, noted, "There are some thirty species of owls (Strigiformes) in Mexico and North America. . . . The fact that English- or Spanish-speaking field workers typically think of 'owl' as . . . homogeneous is not a testimonial to our abstractive abilities, but rather indicates a divorce from our natural surroundings. Native peoples were not so isolated."

An Inuit boy's clothing features a variety of amulets, including an owl claw at the upper right, intended to give him strength. The Inuit people revered the Ookpik (Snowy Owl).

Throughout the surviving Native American languages, owls themselves go by many names, including a variety of monikers for the same species we know today. All these names are phonetic—we infer spellings from the oral language. Many owl names are onomatopoeic, meaning they are derived from the bird's vocalizations: the Cherokee word for Barred Owl is Ubuku, which recalls this owl's iconic "who-cooks-for-you" call, and in the Mayan Tzeltal language, Toy-toy is the pygmy-owl, known for its simple "toot-toot" calls. Owl names could vary among dialects. The Shoshone language, for example, has a variety of

Principe Scops-Owl (*Otus bikegila*)

different names for the same owl. The Burrowing Owl could be called Pookoo, Goto', Kah-tot'-se, Ko'-ho, or Ko'-to-tě, depending on the dialect (and on the transcriber's ability to create a written description of words in nonwritten languages).

The Shoshone lived in the Great Basin and parts of the Northern Rockies, so other owls whose various Shoshone names have survived include the Western Screech-Owl, Great Horned Owl, and Barn Owl. Likely they had names for other species as well, but preserving native languages has always been challenged both by the lack of native speakers and by the ethnographers aiming to preserve words: if a 19th-century linguist had never seen nor was aware of the Northern Saw-whet Owl, or if the surviving speakers of Shoshone being interviewed didn't know of the bird, the native word for that species could simply vanish forever. But among surviving Native languages, owls enjoy an amazingly vast assemblage of colorful and descriptive names.

Best Kept Secrets

Surprisingly, happily, the world of owls still has many secrets to reveal—including whole new species. Just a few years ago, researchers confirmed the existence of a diminutive owl living in a tiny area on a small island off the coast of Central Africa. Príncipe Island is part of the Democratic Republic of São Tomé and Príncipe and covers a mere—but dramatically scenic—52.5 square miles. Within the old-growth tropical forest on the island's south end, the newly described Principe Scops-Owl haunts the night with its characteristic *tuu-tuu-tuu* serenades. Spurred on by local reports dating back nearly a century, scientists began looking for the owl in the late 1990s and finally confirmed its existence in 2016. Intensive study thereafter, including capture and examination of specimens, systematic surveys of the island, and more, finally revealed—in 2022—that the owl was indeed a new species.

For the owl's Latin name, the research team chose *Otus bikegila*. The *Otus* genus has dozens of member species; *bikegila* honors a man named Ceciliano do Bom Jesus, known as Bikegila. A native of Príncipe Island, Bikegila is a former parrot harvester who became a warden in the Obô Natural Park of Príncipe soon after its creation in 2006. According to the Principe Scops-Owl research team, Bikegila "began the 'Príncipe Scops-Owl saga' in 1998" by sharing details about a pair of owls he had seen in a parrot nest. "Since then," says the team, "Bikegila took part in every field effort that led to the bird's discovery for science; he also led the capture of all sampled individuals, including the holotype, which required ingenious ways to erect canopy nets. For almost 25 years, Bikegila has put all his resources, including bottomless fieldwork skills and a vast knowledge of Príncipe, towards the successful completion of innumerable research projects" that took place in brutally inhospitable terrain.

The confirmation of this little owl's existence follows on the heels of several other such discoveries over the past few decades. With the dynamic nature of owl taxonomy, scientists continue to refine the genetic relationships between species. Sometimes the work leads to recognition of new species and sometimes to the realization that owls once thought to be different species are subspecies. But the taxonomy, coupled with the challenge of studying secretive creatures that are active at night, often in remote places, makes for compelling stories, and from these efforts come opportunities to save extremely rare owls such as the Principe Scops-Owl, which is critically endangered owing to very limited habitat and low population estimates.

Titanic Terror!

The familiar Great Horned Owl is a large bird, about the same size as the equally well-known Red-tailed Hawk. But both these ferocious predators are mere Lilliputians compared to the largest owl the world has ever known. Imagine an owl nearly four feet tall, and weighing upwards of 30 pounds, with massive, powerful legs and gigantic feet equipped with daggerlike talons; envision a huge, sharp bill capable of tearing flesh to shreds. Sounds like a feathered ogre from some horrifying netherworld, but this monstrous owl was very real. Meet *Ornimegalonyx*, the extinct Cuban Giant Owl.

When subfossils (bones slowly becoming fossils) were discovered in Cuba in 1954, scientists at first assigned them to the prehistoric bird family Phorusrhacidae—the so-called Terror Birds. However, further examination revealed them to be owl remains, from a giant that hunted and haunted Cuba in the late Pleistocene, which persisted from about 129,000 to about 11,700 years ago. The gargantuan owl was entirely or nearly flightless:

The American Museum of Natural History's life-size model of the extinct Cuban Giant Owl

both gigantism and flightless-ness are common evolutionary occurrences on islands that don't have any endemic large mammalian predators. *Ornimegalonyx* must have been a walking, running, sprinting, pouncing, hissing, screeching menace to virtually every terrestrial creature on its island paradise, but especially to the various species of hutia—large rodents that look like a cross between a rat and a Guinea pig. Half of the nearly two dozen species of hutia are extinct now, but for millennia before humans invaded their world, these rodents were abundant in Cuba and the fearsome *Ornimegalonyx* was their worst nightmare.

We don't know what caused the extinction of the Cuban Giant Owl. Over the span of millions of years, extinction is the rule not the exception, and when humans have landed on previously unpeopled islands, extinctions always follow for various reasons. Hawaii, for example, was once home to four species of largely terrestrial Stilt-Owls, known from subfossils. When Polynesians reached the Hawaiian Islands more than a thousand years ago, these owls went extinct, probably very quickly because they had no natural defense against human hunters, habitat destruction, and the inevitable invasive rats, dogs, and pigs that accompanied these seafaring people. The Cuban Giant Owl, however, may have been extinct by the time humans first reached the island—the scant few *Ornimegalonyx* remains so far recovered date to a time prior to the accepted dates for first human occupation, but we don't have a complete picture of either event—the owl's extinction and the arrival of humans to Cuba—to surmise a causative relationship.

Why Build When You Can Buy (or Steal)?

Apparently, owls are too busy with field sports (aka hunting) to bother with hard labor, so when it comes to home-building, they want nothing to do with it. Far easier to simply appropriate what you want (if you're a big owl), or what you can (if you're a little owl). Unlike most birds, owls don't build nests. In fact, the closest thing to nest-building in the world of owls are the modest *scrapes* made by ground-nesters such as the Snowy Owl and the Short-eared Owls. These species typically go to the trouble of scratching out shallow depressions on the ground, and even that modest effort is too much for the other owls. Most owls nest in cavities and crevices, often in trees (especially dead or dying trees), frequently using holes excavated by woodpeckers.

But big owls can't squeeze into most woodpecker holes, so many species rely on larger hollows and crevices, in both trees and on cliffs or similar places, and in some cases (especially with Barn Owls) in structures built by humans. Many small and medium-size owls will readily use artificial nest boxes created for them; even the large Barred Owl, so familiar in the eastern half of the United States, can be lured to nest boxes built big enough to accommodate them. But the largest owl species are sometimes called platform nesters because they can make use of just about any more-or-less flat surface that supports them, their eggs, and then their chicks. The Great Gray Owl, for example, often nests in the top of large, broken tree trunks or in the broad fork where a big limb meets a tree trunk. But more often, like their robust counterpart the Great Horned Owl, these birds use stick nests built by other species, such as hawks or ravens.

Many large owls are platform nesters, able to use a wide array of flattish surfaces for nests. Here, Great Horned Owls have nested on the broken trunk of a large saguaro cactus in Arizona.

As they often do, a Great Horned Owl took possession of a Great Blue Heron nest in the center of this heron rookery in Colorado. Most owls don't build or even make improvements to nests, and large species often commandeer stick nests made by other birds and even mammals, such as squirrels.

Great Horned Owls, in fact, use a wide array of stick nests. They are only too happy to appropriate a big stick nest built by a heron—a comparative mansion in the world of bird nests—and often take up residence in nests built by Red-tailed Hawks. But this kingpin of North American owls is highly adaptable: Great Horned Owls, which occupy virtually every habitat in the United States, lay eggs just about anywhere. In 1938, after a review of scientific reports on the species, ornithologist Frederick M. Baumgartner realized "that the choice of nesting sites of the Great Horned Owls throughout their wide range includes almost every type of situation in which birds nest, a range of variation unequalled by any other North American bird. From extreme heights of almost a hundred feet to badger and coyote dens in the ground, the situations include old nests of other birds, hollow trees and stumps, holes and ledges on cliffs, and even the open ground."

Blinded by the Light

"The owl seeth by night . . . in the day time she is half blind," we are informed by a 17th-century textbook. Such was the wisdom of the day, but despite persistent ages-old folklore to the contrary, owls are not blinded by daylight. However, light certainly plays an important role for nocturnal hunters and the hunted, and Barn Owls, which inhabit much of the world, use moonlit nights to their advantage. Depending on the subspecies, Barn Owls range from nearly pure white to rich tawny shades ("red" Barn Owls) on their undersides, and scientists have long wondered about the differences. In a unique series of experiments, researchers in Switzerland studied a population of Barn Owls for 20 years and determined that red Barn Owls are less successful at capturing prey during moonlit nights, but white Barn Owls do just fine under the same conditions.

Less successful hunting on moonlit nights led to less successful brood production for the red Barn Owls if their eggs hatched during periods of bright moon. But why the difference? Why did the white Barn Owls gain a predatory advantage under the light of the moon? The key, the team discovered, was in the way voles (small mouselike rodents and the primary food for Barn Owls) reacted to the sudden appearance of either a predator or a bright light. Their first instinct is to freeze, to not move a muscle, so that perhaps they will remain undetected until the threat has passed.

Barn Owl

Boreal Owl

Experiments demonstrated that with the aid of moonlight, voles detected both red and white Barn Owls better than on dark, moonless nights, but also—intriguingly—that the little mammals froze for longer periods of time when white Barn Owls approached. Staying put longer, hoping not to be noticed, however, the voles instead put themselves at increased risk of discovery and sure enough, white Barn Owls caught more voles on moonlit nights than red Barn Owls.

Moreover, further investigation determined that white Barn Owls actually use their bright plumage to gain further advantage—they often approach prey with the moon in front of them, making their undersides like a big light reflector. The sudden appearance of such a bright surface startles the voles in a way that red Barn Owls can't match—and this interaction provides insights into the evolutionary nature of color in nocturnal animals.

Welcoming Wails

As we have seen, owls are wonderful, fascinating, enigmatic creatures. Yet in many parts of the world, owls still endure widespread persecution wrought by misunderstanding, and even in North America they've needed centuries to claw their way past folklore that villainized them. When Hilda Roberts compiled more than 1500 superstitious beliefs from Louisiana for her master's thesis at the State University of Iowa in 1923, owls were abundantly represented. Unsurprisingly many of these superstitions connect owls with doom, gloom and death: an owl hooting when the moon is shining is a sign of death; if an owl hoots from a tree near your house, someone is not long for this world; if you hear an owl hoot three times, a family member will die; if a screech-owl hoots near your window or has the audacity to fly into the house and perch on a bed, a death is assured. Killing an owl was considered bad luck, so, given the baleful nature of these evildoing birds, it's no wonder these owl-fearing Louisianans curated myriad solutions to hearing or seeing an owl, hoping to ward off cataclysmic events. For example, "If an owl hoots, turn the tongue of your shoe inside out and spit between the first two fingers of your right hand, and the owl will not harm you," or to make a screech-owl leave, tie a knot in the corner of your apron, or to make the little owl stop screeching, "take off your coat and turn one sleeve wrong side out."

The Americas, of course, were ripe with owl mythology long before Europeans arrived, and while Roberts thought the Louisiana superstitious beliefs were largely derived from Europe, owls have been both loathed and lionized all over the world throughout human history. They are easy targets for prejudice because they are inherently difficult to understand given their secretive, nocturnal lifestyles, but we are moving beyond old falsehoods, and the world over, many peoples have learned to adore and appreciate owls, as the ancient Greeks did when they associated the Little Owl with Athena, the goddess of wisdom and reason. So, if indeed an owl hoots three times or screeches near your open window, count your blessings, for you've just been invited into a journey of discovery.

Owls of the United States and Canada

Tim braked suddenly, shifted the truck into reverse, and backed up a few yards. My coauthor for *Birds of the Pacific Northwest*, and an expert birder, had spotted something on the hillside above us, but the first stars were twinkling through the fading twilight and all we could make out was a silhouette atop a wooden fencepost. Bending low at the waist and craning my neck so I could see through Tim's driver's side window, I recognized the shape of a small owl just 20 yards distant.

"Western Screech-Owl?" Tim mused.

"I guess so, but I'm not seeing any ear tufts," I replied. Tim suggested the bird must have had its oft-prominent ear tufts lowered, making its head appear smooth and rounded. Through the faded gloaming

Great Horned Owl

we couldn't discern patterns or colors. We sat there with the truck idling; the owl stayed put, and we began to question our identification. We could just make out each shift and turn of the owl's head, and still no ear tufts.

The gravel road coursed along the Oregon side of Brownlee Reservoir, an impoundment that forms part of the border with Idaho and which partway fills a massive cleave in the earth that was once home to a mighty free-flowing Snake River. The miles-long canyon is so deep that even the impounded Snake River can barely fill the bottom, and the rugged uplands towering above are dominated by sagebrush steppe, with steep slot canyons that carry tributary rivulets to the reservoir. These canyons provide luxuriant riparian stands of willow, chokecherry, and cottonwood—perfect habitat for Western Screech-Owls. Hence our initial identification of the enigmatic little owl perched on the fencepost above us on a warm late-spring evening.

Aloud, we processed the possibilities—what small owl without ear tufts could be found here? Northern Pygmy-Owl? Possible but not likely; generally they are daylight hunters, and besides, our bird appeared more robust than a pygmy. Saw-whet Owl? Highly unlikely many miles distant from conifer forest, and a Saw-whet would be smaller than the bird on the fencepost. Our bird was too small and compact to be a Short-eared Owl or Barn Owl; the nearest Boreal Owls are seen far to the north. Perhaps a Burrowing Owl had wandered into the canyon from the south.

We were on the verge of convincing ourselves that our mystery owl was a Burrowing Owl, when suddenly a second owl—clearly a Western Screech-Owl, with its prominent ear tufts easily discernible in silhouette—fluttered down to the fencepost and fed some unlucky creature to its fledgling, which had been waiting patiently all the while. Tim and I laughed at ourselves for not remembering that a juvenile screech-owl wouldn't yet have well-developed ear tufts.

That encounter underscored the fact that owls can present identification challenges. However, frequently with owls, identification is easy, at least once you familiarize yourself with the species likely to be found where you find yourself. And that's a big part of both owl identification and bird identification in general: familiarize yourself with ranges to eliminate the unlikely. Range refers to the area in which a species lives, and ranges can be very broad (the Great Horned Owl is found virtually throughout North America), or highly restricted (north of Mexico, the Whiskered Screech-Owl is found only in extreme southeast Arizona). So, you can expect to find a Great Horned Owl just about anywhere, but you'll need to visit that tiny little slice of Arizona to see a Whiskered Screech-Owl in the United States.

Juvenile Western Screech-Owl

Habitat within an owl's range is your next consideration when you are trying to identify an owl. Some owls are habitat generalists to varying degrees, and others thrive only in specific types of habitats. For example, the Barred Owl and Spotted Owl are similar in appearance and in some places their ranges overlap. However, Barred Owls are more general in their habitat requirements than Spotted Owls. So, for example, if you see a medium-large dark-toned owl without ear tufts in the Pacific Northwest, consider the habitat: if your owl is in a big city park or a lowland grove of mixed hardwoods, odds are that you've found a Barred Owl. But if you've found this owl in the mountains, deep within old-growth conifer forest, take a closer look because you may well be enjoying the increasingly rare opportunity to see a Northern Spotted Owl (but even then, double check the field marks—the colors and plumage patterns—because, as habitat generalists, Barred Owls can and do occupy the same habitat required by Spotted Owls).

With a sense of what owl species can be expected in which places, you have narrowed the possibilities. Beyond that, identification comes down to field marks—the observable characteristics that distinguish one owl species from another, as well as behaviors, and vocaliza-tions. Behaviors can provide valuable clues: did you see a pale-colored owl coursing low over open country during daylight? You might be tempted to think Barn Owl—they are found throughout most of the Lower 48—but they are also almost entirely nocturnal. The widespread Short-eared Owl, however, is pale colored and tends to glide and flap over open country to hunt rodents; they frequently hunt during the crepuscular periods (dawn and dusk) and even during daylight, and their buoyant flight pattern (described as mothlike) is characteristic.

Identification by sound is probably more applicable to owls than any other families of birds. Most species of owls are nocturnal and/or crepuscular, and secretive during the day. When you can't see them, you may still hear them, and most owls in North America have distinctive calls. Learning to identify owls by their calls is great fun, whether it's the deep familiar hooting of the Great Horned Owl, the eerie shrieks of the Barn Owl, or the exotic calls of the Barred Owl.

This identification guide to the owls of the United States and Canada provides details on all aspects of identifying owls: range (including maps) and habitat, field marks and vocalizations, behaviors, information about nesting habits and migration habits, and even status details that relate the relative abundance of each species. The status details are gleaned largely from the International Union for Conservation of Nature Red List of Threatened Species and from the American Bird Conservancy population estimates.

Barn Owl

(*TYTO ALBA*)

Size: Medium. Length: 14–16 inches. Wingspan: 40–48 inches. Weight: 14–24 ounces.

With its extensive worldwide range, proclivity for cohabitating with humans, haunting calls, and elegant beauty, the Barn Owl is one of the best-known owls around the globe. It is also among the most studied, and yet in some ways one of the most enigmatic. That's because the genus *Tyto* includes several species and myriad subspecies, and little is known about the life histories of many of them. In fact, the taxonomy of the Barn Owls is likely not yet

Owls of North America

settled. For decades, *Tyto alba* comprised most Barn Owls worldwide, but then the genus was divided into several species, with a major split being new full species status for the American Barn Owl (*T. furcata*) and its dozen or so subspecies found throughout the Americas. But that decision was later reversed, and most organizations are now back to recognizing *T. alba* as the single species of Barn Owl. Its two-dozen-plus subspecies range from the relatively pale-colored Barn Owls of North America to much darker subspecies like the Lesser Antilles Barn Owl.

Identification: Tall and slender, with white underparts that are often lightly peppered in tan; soft, rich, golden-tan upperparts surround a gray mantle that is sparsely flecked with small white spots. White, heart-shaped facial disk is clearly delineated by a brow of dark feathers. Seen from below in flight, especially near lights at night, Barn Owls

Barn Owl

Barn Owl

Barn Owl

often appear snowy white; all the flight feathers have dark chevrons that align to form pronounced wing bars, much more prominent on the dorsal wing surface than the ventral wing surface; day-flying Barn Owls are rare unless they have been disturbed and flushed from a roost. *Fledgling:* Like adult but typically paler.

Voice: Common call is a loud, high, raspy shriek that can raise the hairs on the back of your neck; males shriek more often than females, and both sexes use a softer, less harsh version of this shriek, as well as a somewhat musical cricketlike twitter, and a soft but persistent hiss. Agitated Barn Owls make a clicking noise by snapping their bills.

Habitat: Barn Owls use a wide range of habitats, generally preferring open areas for hunting, but with diversity of cover options; within the United States they generally avoid dense forest and high mountains. Australian researcher Murray D. Bruce suggests that the Barn Owl's habitat is "probably limited only by prey availability, by seasonality and the severity of winters at higher latitudes."

Range: In the Americas, inhabits the southern third of North America, along with Central and South America, and the Caribbean islands. Northernmost range is southern British Columbia; sparse and spotty distribution from central Idaho eastward across Montana, Wyoming, the northern Great Plains, Great Lakes states, West Virginia. Absent from northernmost Great Plains and northern New England. Generally absent from high mountains in the United States. Also found on all Hawaiian Islands after being introduced there by the Hawaii Board of Agriculture and Forestry between 1958 and 1963 to control populations of non-native rodents.

Migration: Northernmost populations can be partly migratory, withdrawing southward for the winter.

Status: Widespread but generally uncommon; rare in the northern fringes of its range. North American population 140,000. IUCN Least Concern, but endangered and declining in Canada and northern states.

Range of Barn Owl

Diet: Small mammals (mice, voles, and many others) are the primary prey of Barn Owls, but they are opportunistic when necessary and will sometimes take birds, insects, rabbits, and even bats.

Nesting: Barn Owls nest in cavities and crevices, ranging from tree holes and nest boxes to cracks and cavities in cliffs; they adapt readily to human activity and nest within buildings and other structures, haybale stacks, bridgeworks, and more; they are capable of excavating dens in soft soils of dirt cliffs and slopes.

Clutch: Four to five (sometimes up to 12), 1.6-inch-long whitish eggs.

Behaviors: Barn Owls are primarily active at night and have excellent night vision but also superb hearing with which to locate and capture prey. Their long legs are well adapted to plunging through snow cover and deep grass to snatch small mammals they've located by sound. They primarily hunt in flight, coursing low over open landscapes.

Similar species: Because the Barn Owl's ventral (front) side is white, the bird is sometimes confused with the Snowy Owl, usually by inexperienced or casual observers seeing the bird in headlights or similar circumstances. Barn Owls are much smaller and slimmer than Snowy Owls, and the range of the two species only overlaps in the southernmost wintering regions for Snowy Owls. The Barn Owl can also be confused with the similar-size Short-eared Owl, and both species often hunt in open country, but Short-eared Owls often hunt in daylight and Barn Owls rarely hunt before dusk or after dawn. Short-eared Owls are darker overall, with bold, dark streaking. Among North American owls, only the Barn Owl has the white, well-defined, heart-shaped facial disk.

Barred Owl

(STRIX VARIA)

Size: Large. **Length:** 17–21 inches. **Wingspan:** 40–44 inches. **Weight:** 1.25–2 pounds (females 1.3 to 1.4 times heavier than males).

With its familiar call and expansive range, the Barred Owl has been one of the most familiar owls east of the Great Plains for hundreds of years, but in the 20th century this bird completed a significant range expansion all the way to the West Coast—with dire consequences for its close relative, the Northern Spotted Owl. Despite its stature as the fourth-largest owl in North America, the Barred Owl—like almost every other owl on the continent—is subject to predation by the notorious Great Horned Owl and will vacate its territory if its slightly larger nemesis invades.

Identification: Large, with rounded head, no ear tufts, and all-dark eyes; mottled brown, light gray, white/pale tan, with dark vertical streaks on white flanks and lower breast, bordered by horizontal streaks on throat and upper breast (Spotted Owl has white horizontally oblong spots on a dark background on the breast and flanks); dorsal surface patterned with large brown and white spots; light-colored facial disk. *Fledgling:* Like adult but paler.

Barred Owl

Barred Owl

Barred Owl

Barred Owl fledglings

Range of Barred Owl

Voice: Iconic and oft-heard song is a deep, rich, hooting, mnemonically described as "who cooks for you" and often preceded by a series of ascending hoots increasing in urgency, *who-who-who-who-cooks-for-you*. The final hoot frequently wavers: *who-who-who-who-cooks-for-yoouuu*. Also, a variety of screeches and wails.

Habitat: Mature deciduous, mixed, and coniferous woodlands, including timbered swamps, montane slopes, and even large, forested parks and similar sites within rural and urban areas.

Range: Eastern half of the United States and adjacent southeastern Canada, northwestward across a narrow swath of central Canada; in the West, southeast Alaska southward through British Columbia, much of Alberta, Washington, Montana, Idaho, Oregon, and Northern California; disjunct population in parts of central Mexico.

Migration: Nonmigratory.

Status: Uncommon to locally common. Population 3.2 million and stable or increasing. IUCN Least Concern.

Diet: Small mammals, birds, insects and other invertebrates, reptiles and amphibians, occasionally even fish and crayfish.

Nesting: Primarily uses natural cavities and crevices in trees, but also stick nests built by other raptors or Corvids (e.g., ravens and crows); readily uses artificial nesting boxes.

Clutch: Two to four, 2-inch-long white eggs.

Behaviors: Perch and pounce predators, Barred Owls sit on branches under the forest canopy and listen and watch for prey. They will drive off, kill, and sometimes hybridize with Northern Spotted Owls.

Similar species: See Spotted Owl. Great Gray Owl has yellow eyes and a large, well-defined facial disk.

Boreal Owl

(*AEGOLIUS FUNEREUS*)

Size: Small. Length: 8–11 inches. Wingspan: 22–25 inches. Weight: 3–7 ounces.

Closely related to the more southerly Northern Saw-whet Owl, the striking little Boreal Owl is a forest dweller of the north, its circumpolar range reaching nearly to the arctic circle. Owlers hoping to see this species face a triumvirate of problems: Boreal Owls are small, nocturnal, and quiet outside of their breeding season. That's why owl specialists listen for them in the late-winter/early-spring courtship and breeding season, when Boreal Owls utter their distinctive series of speedy toots in quick succession. French fur traders gave this owl "the liltingly beautiful French-Canadian name *la nyctale boreale*—the night owl of the north," explains 19th-century ornithologist Clinton Hart Merriam.

Identification: Small owl with squarish head that often appears overly large for the body size; bright yellow eyes set in a well-defined pale gray facial disk framed by dark brown with varying degrees of white flecks; crown patterned with white spots; bill ranges from yellowish to whitish to gray. *Fledgling:* Overall dark brown, with paler smudges on the breast/flanks, and pale, poorly-defined eyebrow and mustache streaks.

Voice: Courtship-season song is a series of 8 to 20 short, sharp, trilled toots in quick succession; the male's singing can go on for several hours; calls include raspy chirps and sharp haunting wails.

Habitat: Boreal forest and subalpine montane forest.

Range: Year-round range extends from Alaska across most of subarctic Canada and into the Pacific Northwest, and south through the Rocky Mountains to northern New Mexico.

Migration: Nonmigratory or partially migratory, but in winter, nomadic and possibly occasionally irruptive.

Status: Uncommon to locally common. Global population 730,000 to 1.8 million mature birds. IUCN Stable.

Boreal Owl

Boreal Owl

Range of Boreal Owl

Boreal Owl fledglings

Diet: Primarily small mammals, such as voles, mice, flying squirrels, shrews; also small birds and insects.

Nesting: Cavity nester; uses woodpecker holes and natural cavities.

Clutch: Three to six, 1.3-inch-long dull-white eggs.

Behaviors: Perch and pounce hunter; mostly nocturnal except in far north during season of extended daylight hours, when it must hunt diurnally. In the 1980s, a researcher discovered that during the late-winter/spring mating season, male Boreal Owls will sing from up to five different potential nest cavities and surmised that "the number and quality of nest holes defended by males may have been of particular importance" to females in choosing a mate.

Similar species: Smaller Northern Saw-whet Owl has rust-and-gray plumage, its facial disk is less defined, crown is streaked rather than spotted, and bill is black rather than pale.

Burrowing Owl

(*ATHENE CUNICULARIA*)

Size: Medium. Length: 8–10 inches. Wingspan: 21 inches. Weight: 5–8 ounces.

Unique among owls, the iconic Burrowing Owl, with its bemused, almost comical appearance, nests underground by commandeering abandoned tunnels dug by mammals such as ground squirrels, prairie dogs, and even badgers. If need be, these owls of wide-open spaces can renovate their upcycled homes and even dig their own—in fact, the Burrowing Owls of Florida and the Caribbean islands, where burrowing creatures are rare, often excavate their own homes. Found from Canada to southern South America, Burrowing Owls comprise 18 subspecies, two of which are extinct. The Western Burrowing Owl (*A.c. hypugaea*) is found from Canada to Central America, while the Florida Burrowing Owl (*A.c. floridana*) occupies its namesake state and the Bahamas.

Burrowing Owl

Identification: Note penchant for daylight perching on the ground near nesting burrows and on tall nearby objects such as fenceposts. Long legs, upright posture, yellow eyes, prominent white eyebrows, rounded head; frequently bobs up and down, especially in presence of humans or other intruders. *Fledgling:* Pale version of the adult and typically seen standing on the ground near burrow.

Voice: A short, sharp, two- or three-note wail is reminiscent of a California Quail; also a variety of rasps, screeches, rattles, and more.

Habitat: Open country, ranging from prairies, steppe, grasslands, and desert to human-modified locales such farmlands, ranchlands, airports, roadsides, and more.

Burrowing Owl

Adult Burrowing Owl (left) with four juveniles

Range of Burrowing Owl

Western Burrowing Owls tend to be most numerous where burrowing mammals are most numerous.

Range: In North America, Burrowing Owls range from south-central Canada to Central America, and from the Great Plains westward to eastern Washington and eastern Oregon, and south through much of California, the Desert Southwest, and Texas; they are rare and local along the Gulf Coast from Louisiana to the Florida Panhandle and widely distributed throughout the remainder of Florida.

Migration: Throughout most of their range in south-central Canada and the United States, Burrowing Owls are migratory. Especially in California, but also across the southern halves of Arizona and New Mexico, and through much of Texas, some Burrowing Owls remain year-round (and a few remain year-round in the northern portions of their breeding range). Wintering range extends southward into Central America. Disjunct populations in Florida and the Caribbean (as well as South America) are nonmigratory.

Status: Rare to locally common. Population unknown and declining, and listed as endangered in parts of Canada and some states. IUCN Least Concern.

Diet: Burrowing Owls eat a wide range of prey, including insects and other arthropods, small mammals, reptiles, and small birds.

Nesting: Unique among birds of prey, Burrowing Owls nest underground by appropriating abandoned (or soon to be abandoned) burrows dug by mammals such as ground squirrels, prairie dogs, kangaroo rats, tortoises, skunks, badgers, armadillos, and marmots. However, Burrowing Owls in Florida and the Caribbean dig their own burrows owing to a lack of burrowing animals.

Clutch: Up to 12, 1.25-inch-long white eggs.

Behaviors: Burrowing Owls are frequently active during daylight, particularly morning and evening, but also at midday unless temperatures soar. They are adept at hunting on the ground, walking, scampering, and pouncing to snatch up grasshoppers and other small prey, and they can also hawk insects from the air, as well as glide and hover in search of food.

Similar species: See Short-eared Owl, which sometimes roosts on the ground and sometimes hunts in daylight.

Eastern Screech-Owl

(*MEGASCOPS ASIO*)

Size: Small. Length: 7.5–9 inches. Wingspan: 20–22 inches. Weight: 5–7 ounces.

One of the best-known and most beloved owls in the United States, the Eastern Screech-Owl occurs in many different types of treed habitats, from wilderness woodlands to urban sites such as parks, riparian strips, cemeteries, campuses, and even yards. They adapt so well to the trappings of humanity that they routinely thrive in settled areas, from farm- and ranchlands to favorite wooded patches even in the largest cities. Each of five subspecies occupies its own geographic area within the species' range, with significant overlap.

Identification: Overall richly mottled gray/black/white (*gray morph*) or rust/black/white (*red morph*) or intermediate gray-brown; prominent black crosshatch marks down the breast and flanks. Prominent ear tufts usually visible, but can be lowered; stocky appearance,

Eastern Screech-Owl, gray morph

Eastern Screech-Owl

Eastern Screech-Owl, rufous morph

Juvenile Eastern Screech-Owl

with large yellow eyes; pale gray or pale greenish bill. Among the five subspecies, red morphs are most common in *M. a. asio* and *M. a. floridanus*, which together occupy the eastern two-thirds of the species' range. Note that the dissimilar Northern Saw-whet Owl is the only small owl to share significant range overlap with the Eastern Screech-Owl (see "Similar species"). *Fledgling:* Heavily patterned with gray/white or rust/white horizontal stripes much like the plumage pattern of a Plymouth Rock chicken.

Voice: An eerie wavering whinny is iconic throughout the range of the Eastern Screech-Owl and is thought to be used to defend territories. They also use a rapid, even bouncing-ball-type song, which can last several seconds, along with haunting screeches and various barking-type calls.

Habitat: Primarily wooded habitats ranging from sparse forest to dense forest thickets in open areas such as fields, meadows, parks; wooded riparian areas; adapts well to human-altered habitats if nesting cavities and roosting sites are available.

Range: Most of the United States east of the Rocky Mountains, into parts of south-central and southeastern Canada, and south into northeastern Mexico.

Migration: Nonmigratory.

Status: Common. Population 500,000 mature birds and declining; IUCN Least Concern.

Diet: Insects and other invertebrates, reptiles, small mammals, birds, and even fish and amphibians; one study of Eastern Screech-Owl diets in the northmost part of their range (in Manitoba) found that invertebrates composed a significant majority of their prey and included earthworms, perhaps because, as the authors postulated, "the watering of lawns at night with sprinkler systems may provide greater access to earthworms in suburbs." Farther south, a study

Range of Eastern Screech-Owl

in Kentucky found these owls to eat lots of invertebrates and lots of small mammals, but also significant numbers of crayfish. In virtually all cases, insects and other invertebrates are the most numerous prey items, but small mammals and birds provide the most biomass of prey.

Nesting: Nests in cavities, including woodpecker holes, natural cavities in tree trunks, and nest boxes built specifically for owls and for other species such as Wood Ducks; because Eastern Screech-Owls have adapted well to humans, they've been known to nest in a variety of human-made cavities: for several years running, for example, a pair of Eastern Screech-Owls nested in an old decorative mailbox at a home in Florida.

Clutch: Two to four, 1.3-inch-long whitish eggs.

Behaviors: Like other cavity-nesting owls, Eastern Screech-Owls also use cavities and cavity openings for roosting, especially in winter.

Similar species: Very similar in appearance to the Western Screech-Owl, but the area of range overlap between the two species is minimal (primarily parts of Texas and Colorado). The calls are different between the two species (see descriptions), and the Western Screech-Owl has a dark grayish-black bill.

Elf Owl

(*MICRATHENE WHITNEYI*)

Size: Small. Length: 5–5.5 inches. Wingspan: 9 inches. Weight: 1.5 ounces.

As its name might suggest, the Elf Owl is the smallest of the owls, or at least tied for smallest with the Long-whiskered Owlet of Peru. In fact (along with its Peruvian distant cousin) this diminutive dynamo is the daintiest raptor of any kind in the world—even smaller than such aptly-named bantamweights as the Tiny Hawk of South America and the Pygmy Falcon of Africa, and ever so slightly smaller than the sparrow-size Black-thighed Falconet of southeast Asia. The Elf Owl, beloved by birders and owlers, thrives in desert ecosystems; a tiny, fierce-looking owl face peeping out of an old woodpecker hole in a saguaro cactus is an iconic image of the Southwest.

Identification: Tiny grayish owl, with varying degrees of rufous, and no ear tufts; breast streaks appear more like smudges on a white background rather than well-organized lines or crosshatches, coalescing into broad lines on the flanks; bright yellow eyes, and angled to gently curved white eyebrows. Outside edges of facial disk are poorly defined. *Fledgling:* Mottled in pale tones of gray.

Elf Owl with prey

Elf Owl

Elf Owl in nesting hole

Voice: Common call is a high, squeaky bark, usually in a series; chattering call is reminiscent of yipping coyote pups; also, an urgent high-pitched yelp and various squeaks.

Habitat: A wide variety of desert scrub and scrub-forest habitats, including riparian canyons up to more than 5000 feet.

Range: In the United States, Elf Owls occupy three disconnected regions in the Southwest: the Sonoran Desert from approximately the Lower Colorado River eastward across the southern half of Arizona and into southwestern New Mexico; the Big Bend region of west Texas; and the Lower Rio Grande River below Falcon Dam, primarily in the Bentsen-Rio Grande Valley.

Migration: Elf Owls migrate from the United States and northern Mexico to the uplands of central and southern Mexico, typically departing by October and returning between February and April. According to the authors of the species account for Cornell University's Birds of the World website, "Only the disjunct Mexican populations are nonmigratory, but the situation could be changing with climatic warming." However, winter sightings in the species' northern breeding range remain rare.

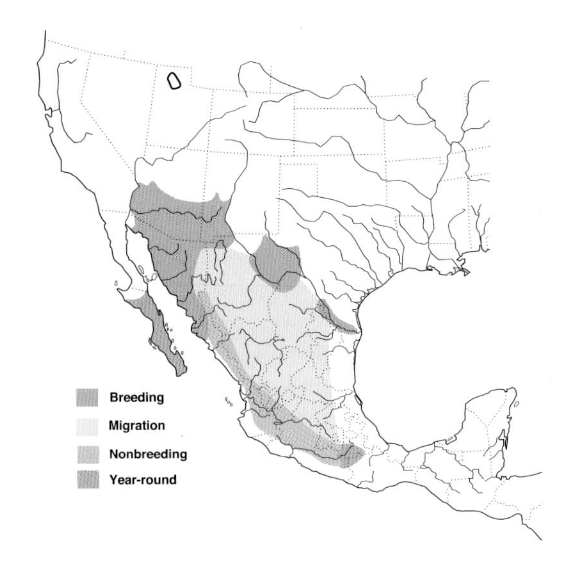

Range of Elf Owl

Status: Uncommon to locally common in the United States. Population 72,000 (U.S. and Mexico) and decreasing. IUCN Least Concern.

Diet: Elf Owls are insectivorous, subsisting primarily on arthropods, including moths and beetles.

Nesting: Cavity nesters, Elf Owls frequently use abandoned woodpecker holes; they use holes made by Gilded Flickers and Gila Woodpeckers in saguaro and other cacti but also use woodpecker cavities in a wide variety of tree species. In trees with numerous cavities, Elf Owls may be among a variety of other cavity-using bird species nesting simultaneously.

Clutch: One to five, 1-inch-long white eggs.

Behaviors: Nocturnal hunters, Elf Owls hunt insects and other arthropods by perch-and-pounce strategy and can hover to snatch bugs from foliage. In a case of turnabout is fair play, Elf Owls (during the breeding season) have been observed mobbing potential predators: researchers in Arizona used a captive Great Horned Owl in their experiments, and when tethered to a perch near Elf Owl nesting locations, the bird was quickly subjected to attack by at least four Elf Owls. Presumably the Great Horned Owl was not amused, and the team reported, "Before we could relocate the horned owl to a protected enclosure, it was struck once in the head by an Elf Owl."

Similar species: Western Screech-Owl is larger, has prominent ear tufts (which can be hidden), lacks the Elf Owl's defined white eyebrows, and has defined dark crosshatched streaks on the flanks and breast as opposed to the Elf Owl's dark smudges; Northern Pygmy-Owl has well-defined dark streaks on the flanks and breast, and a prominent long tail.

Ferruginous Pygmy-Owl

(*GLAUCIDIUM BRASILIANUM*)

Size: Small. Length: 6 inches. Wingspan: 14–15 inches. Weight: 2–2.5 ounces.

Although widespread in Mexico, Central America, and South America, the Ferruginous Pygmy-Owl reaches the northern extent of its range in southern Arizona and southernmost Texas. Unlike the very similar Northern Pygmy-Owl, the Ferruginous Pygmy-Owl is a lowlander, and its range in Arizona is extremely limited, primarily the Altar Valley, which is low-elevation desert rarely inhabited by Northern Pygmy-Owls. Moreover, in both Arizona and Texas, the species was once significantly more widespread but has fallen victim to the rapid expansion of human development that eliminates tracts of mesquite scrub and desert riparian habitat the owl needs for survival.

Identification: Very similar to Northern Pygmy-Owl, but overall rufous toned rather than grayish brown (though some Northern Pygmy-Owls are reddish and some Ferruginous Pygmy-Owls are gray-brown); tail usually brown with lighter-brown to cream bars (white bars in the Northern Pygmy-Owl); fine thin streaks on the crown and forehead. *Fledgling:* Like adult but with less spotting; forehead/crown may be plain cinnamon-brown until spots develop.

Voice: Song is a monotonous high-pitched toot given over and over; also makes a repeated songbird-like two-note *cheerup*, and various other calls near the nest.

Habitat: In Arizona, lives in Sonoran Desert habitat and is likely limited to the Altar Valley; in Texas, occupies mesquite, live oak, and *Pithecellobium* habitats, and riparian areas of the Lower Rio Grande Valley.

Range: Widespread in the Americas, but in the United States only found in Arizona's

Ferruginous Pygmy-Owl

Ferruginous Pygmy-Owl

Altar Valley and the southernmost Gulf Coast region of Texas.

Migration: Nonmigratory.

Status: Rare and local in Texas and Arizona and decreasing; widespread and relatively common in much of Latin America. IUCN Least Concern, but Arizona's very small extant populations face continued decline and potential extinction; in Texas, the species faces similar habitat alteration/destruction but so far has fared better than in Arizona. Total United States population may be fewer than 1000 birds.

Diet: Insects, small birds, small mammals, reptiles.

Nesting: Cavity nester; uses woodpecker holes and natural cavities.

Clutch: Two to seven, 1-inch-long white eggs.

Behaviors: Hunts during the day and during the crepuscular periods and is expert at catching small birds up to about its own size; in 2011, a researcher in Brazil, over the course of three days, watched a Ferruginous Pygmy-Owl snatch four hummingbirds that were feeding at flowers.

Similar species: Northern Pygmy-Owl has white tail bars and fine spots rather than thin streaks on the forehead; Elf Owl has smudges rather than well-defined streaks on the breast and flanks, and has a short tail.

Flammulated Owl

(PSILOSCOPS FLAMMEOLUS)

Size: Small. Length: 6–7 inches. Wingspan: 16 inches. Weight: 1.5–2 ounces.

The diminutive Flammulated Owl is hardly larger than an American Robin and feeds almost entirely on insects, which it hunts after dark in the ponderosa pine forests of the American West. Flams are migratory, wintering from central Mexico southward into Honduras, a life history probably deriving from their insectivorous feeding habit—insects are largely inactive and unavailable during winter in the Western mountains. These tiny, strictly nocturnal little owls—lovingly called Flams or Flammies by birding enthusiasts—are seldom seen, so most "sightings" are actually confirmed by hearing the surprisingly deep hooting calls that commonly emanate from prime habitat during the breeding season.

Flammulated Owl

Identification: The only tiny owl in the United States and Canada with dark eyes, rather than yellow. Overall, handsomely patterned in dark gray with varying degrees of rust shades, and black crosshatched bars down the breast; some individuals are overall more rust-toned than gray and sometimes referred to as *brown phase*. Small ear tufts often flattened, giving the head a blocky shape. Inside top edges of facial disk run diagonally from the upper base of the bill toward the ear tufts to give the appearance of broad, lighter-colored eyebrows. *Fledgling:* Richly patterned in gray speckles, with rusty highlights; note all-black eyes.

Flammulated Owl

Voice: Common call is a surprisingly low-pitched hoot (for such a tiny owl), delivered in a lengthy series, spaced, and as either

Range of Flammulated Owl

single hoots or double hoots. Also, a wide variety of screeches, short wails, and other vocalizations.

Habitat: Montane ponderosa pine forest and mixed forest, usually with ponderosa pine component; typically, somewhat open stands, with denser stands mixed in.

Range: Scattered and noncontiguous range components throughout the West where high-elevation ponderosa-dominated and mixed forest (pine, Douglas fir, other conifers, and associated hardwoods) is common, from Central British Columbia south to California, Arizona, and New Mexico, and eastward from Colorado into Wyoming, western Montana, parts of Idaho; breeding range extends to central Mexico. Winter range reaches southward to Honduras.

Migration: Flammulated Owls across most of the species' range withdraw southward for winter, departing their breeding locales between late August and October, and returning mostly from late April to early May. They are nocturnal migrants.

Status: Locally common to uncommon. Population 37,000 and decreasing. IUCN Least Concern.

Diet: Invertebrates, especially insects, including beetles, moths, crickets, and others.

Nesting: Secondary cavity nester, usually using woodpecker cavities excavated mostly by Northern Flickers, Pileated Woodpeckers, and a few other montane species; occasionally nests in natural cavities. Flams don't build nests in their nest cavities, laying their eggs on whatever is on the floor of the cavity.

Clutch: Two to three, 1.2-inch-long dull off-white eggs.

Behaviors: Flams hunt entirely at night, usually high up in trees. They take insects by perching and watching, and then making short flights to snatch prey from vegetation and from the air; they will also capture prey in understory shrubbery and on or near the ground.

Similar species: All other tiny owls in the United States and Canada have yellow eyes; the Flammulated Owl has all-dark eyes; also note that Flams are largely absent from the United States between late October and early April.

Great Gray Owl

Great Gray Owl

(*STRIX NEBULOSA*)

Size: Large. Length: 24–33 inches. Wingspan: 54–60 inches. Weight: 2–3 pounds.

The Great Gray Owl is one of the world's largest owls—at least in terms of height and wingspan. But its apparent bulk is an illusion. This beautiful, imposing owl of northern forests is a relative lightweight, averaging about 2.25 pounds—more than a pound lighter than the average weight of the slightly smaller Great Horned Owl. The dense plumage that conceals the relative slightness of the Great Gray Owl aids not only in stealthy flight but also provides insulation against the cold of the boreal and montane forests where this species thrives year-round. These mesmerizing owls adapt readily to artificial nesting platforms, which have become important implements in Great Gray Owl conservation strategies in some places.

Identification: Very large and tall; overall mottled in rich, bright gray/white typically mixed with soft brown tones on the dorsal surface; relatively small bright yellow eyes set in large, circular, well-defined facial disk; no ear tufts; white collarlike throat bar to varying extent. *Fledgling:* Like adult but paler, with poorly defined facial disk.

Voice: Characteristic call is a series of deep, resonant hoots, and sometimes double note

Great Gray Owl

Great Gray Owl

Great Gray Owl fledgling

hoots; male's hooting is deeper than female's; other vocalizations include low, raspy yelps and higher-pitched chattering calls.

Habitat: Primarily coniferous and boreal forest, usually with openings such as meadows and bogs; sometimes mixed forest and hardwood forest within or adjacent to montane conifer forest.

Range: Northern latitudes worldwide. In North America, from Alaska south and southeastward through much of subarctic Canada; in the continental United States, spotty distribution includes the Northern Rockies, Cascade Mountains, and Blue Mountains (Oregon/Washington), Warner Mountains (Oregon), and the northern Great Lakes region (northern Wisconsin, Minnesota, Michigan).

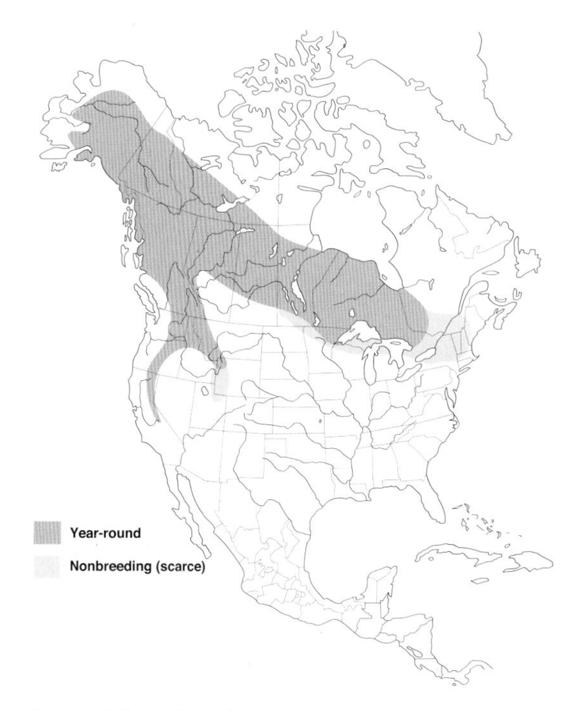

Range of Great Gray Owl

Migration: Nonmigratory, but can be nomadic during winter, extending nonbreeding-season range southward in the Rockies and Great Lakes region.

Status: Uncommon. Global population 50,000 to 100,000 and possibly increasing; North American population 50,000 mature birds. IUCN Least Concern.

Diet: Primarily small mammals, especially voles, but also pocket gophers, mice, and shrews; occasionally somewhat larger prey.

Nesting: Uses stick nests built by other species, hollows formed by broken treetops, artificial nest platforms.

Clutch: Two to five, 2-inch-long whitish eggs.

Behaviors: Great Gray Owls hunt nocturnally and diurnally, though daylight hunting is usually (but not always) near dawn and dusk. They are well-adapted to silent flight, and their superb hearing allows them to precisely locate their preferred prey—voles and other small mammals—then glide in for the kill, even if their victim is well beneath the would-be cover of snow.

Similar species: Barred and Spotted Owls are smaller and have black rather than yellow eyes.

Great Horned Owl

(BUBO VIRGINIANUS)

Size: Large. Length: 18–25 inches. Wingspan: 36–60 inches. Weight: 2.7–5 pounds.

In the Americas, only the Barn Owl ranges over a vaster territory than the majestic Great Horned Owl. Perhaps the most iconic and familiar of all the owls, this fierce, robust predator occupies virtually every habitat type from sea level to high in the mountains. They are one of the largest owls in the world and capable of killing

Great Horned Owl

Great Horned Owl fledgling

Great Horned Owl

Great Horned Owl

Great Horned Owl

and eating animals that substantially outweigh them. Even other large owls, such as Great Gray, Barred, and Spotted Owls, are not safe when Great Horned Owls are on the prowl.

Identification: Large and robust, with long, prominent ear tufts; overall grayish-brown patterned with extensive black mottling, and ranging from overall pale gray to rich dark gray and rust; breast, belly, and flanks densely patterned with horizontal bars, yielding to dark blotches on upper breast, and usually a white throat. Plumage tone varies geographically and seems to correspond to regional climate: the more humid the climate, the darker the owls. Great Horned Owls in the Pacific Northwest are darkest, while those in the Desert Southwest are pale, but the palest birds live in subarctic Canada. Yellow eyes set in gray to rust-colored facial disk that is framed by black on both sides. Large and broad-winged in flight. *Fledgling:* Like a pale version of the adult but lacking prominent ear tufts.

Voice: Deep, soft hooting is the territorial call, most frequently used in winter and spring, and highly variable; typically, *whoo-hoohoo . . . whoo-whoo* or similar; males and females often hoot in duets, with the male's hoots deeper pitched. Other vocalizations are legion, including screams, screeches, whistles, hisses, and various wavering calls.

Habitat: Virtually any partially to predominantly wooded habitat, especially areas with a mix of trees and clearings, ranging from sparsely wooded riparian strips and juniper groves in deserts to deciduous-dominated lowland forests to western montane conifer and mixed forests; also rural to urban areas with sufficient cover for nesting/roosting and open spaces for hunting.

Range: North-central Alaska south to the southern tip of South America, including most of subarctic Canada, all of the United States, and most of Mexico; absent from the Caribbean islands.

Range of Great Horned Owl

Migration: Nonmigratory.

Status: Common. North American population 3.9 million and possibly decreasing. IUCN Least Concern.

Diet: Eats everything from invertebrates to mammals as large as rabbits, hares, and skunks; readily takes other birds, including waterfowl, seabirds, gamebirds, wading birds as large as herons, and other owls; will raid nests of other large birds, such as hawks, at night to capture nestlings, and also takes reptiles and just about anything else.

Nesting: Often uses stick-type nests built by other birds, such as hawks, herons, ravens, and magpies, and sometimes lines the nest with vegetation, feathers, or other materials; nest sites are extremely varied, from high in tall trees to crevices in cliffs to a host of human-made structures.

Clutch: Two to four, 2.2-inch-long whitish eggs.

Behaviors: Great Horned Owls are so formidable that their presence at a specific location often leads other owl species to abandon the area or face the threat of death and consumption; in fact, Great Horned Owls will even kill and eat their own kind; they sometimes attack human researchers that approach nests. They hunt primarily from dusk to dawn but occasionally hunt during daylight.

Similar species: Long-eared Owl is smaller and slimmer; Barred and Spotted Owls are somewhat smaller, lack ear tufts, and have black rather than yellow eyes.

Long-eared Owl

(ASIO OTUS)

Size: Medium. **Length:** 14–16 inches. **Wingspan:** 35–39 inches. **Weight:** 8–15 ounces.

If not for the iconic Great Horned Owl, the Long-eared Owl might vie for the title of quintessential owl, with its regal upright posture, long ear tufts, and piercing yellow eyes set within a rich tawny facial disk. The Long-eared Owl is a denizen of open and semi-open spaces, and ranges across much of North America. I've found them roosting in some remarkable places: a low rimrock rising above a desert playa in Oregon, a tiny box canyon carved into the Nevada highlands, a copse of scrub trees amid Idaho sand dunes, even an old abandoned well house left from pioneer days in the Great Basin. During winter, these birds sometimes roost communally in a favorite grove of trees.

Identification: Often appears tall and slender when perched, with long, black-fronted

Long-eared Owl

Long-eared Owl

ear tufts; light tan to tawny-colored, elongated half-moon-shape facial disk, and bright yellow eyes; flanks, belly, and lower breast patterned with dark crosshatches, with upper breast and throat patterned with irregular dark blotches; sharply angled white eyebrows form a V extending up from the base of the bill. In flight, tawny patch on the outer wing, at the base of the primaries, is less contrasting than the light tan patch on the Short-eared Owl's outer wing. Western birds tend to be paler, and the palest (grayest) specimens I've seen breed in the deserts of the Northern Great Basin.

Long-eared Owl

Long-eared Owl fledgling

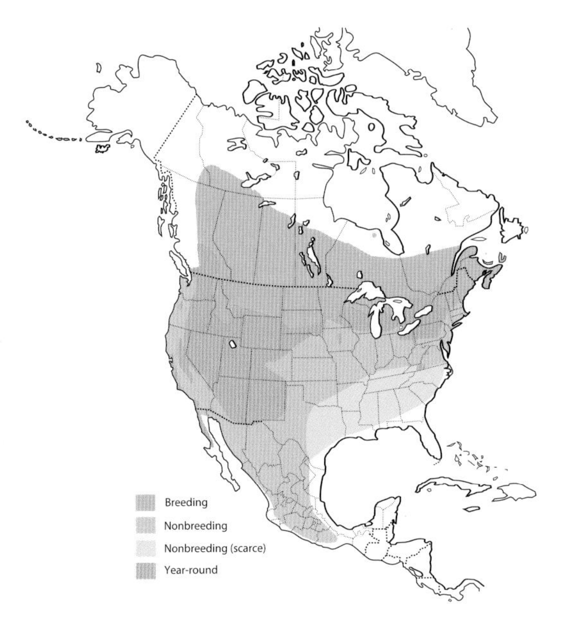

Range of Long-eared Owl

Fledgling: Like adult but paler and more grayish overall, and usually with black surrounding the eyes.

Voice: Deep hoots are common, including the male's lengthy series of extended hoots, each spaced by several seconds, *whooo* . . . *whooo* . . . *whooo*. . . . Also, eerie screeches and wails, and squeaky high-pitched barks.

Habitat: Open country, including scrub-steppe, prairies, farm- and ranchlands, broad meadows within sparse forest, and high desert; hunts over open habitats but prefers to roost in dense vegetation, including riparian corridors, or in cliffs of all sizes, tight against tree trunks, and sometimes on the ground.

Range: Expansive range across the Old and New Worlds; in North America, found coast to coast in the United States and nearly so in approximately the southern half of Canada, absent only from most of western British Columbia; breeding range extends northward to central latitudes in Canada and winter range reaches southern Mexico.

Migration: Largely migrates southward from northern portion of breeding range, though some birds remain there year-round; southward migration can vary from short-distance movements to substantial latitudinal migration, including from Canada and the northern states to Mexico.

Status: Uncommon. North American population 140,000; global population 2.2 to 3.3 million and decreasing. IUCN Least Concern.

Diet: Mainly small rodents, but occasionally songbirds, reptiles, shrews, and even bats.

Nesting: Long-eared Owls repurpose old stick nests built in trees, sometimes cliffs, by hawks, magpies, crows, ravens, and other such species; in Nevada, I once found a pair repurposing a wood rat midden in a low cliff. They make no improvements to the existing nest.

Clutch: Two to ten, 1.6-inch-long white eggs.

Behaviors: Like the closely related Short-eared Owl, Long-eared Owls are agile, acrobatic fliers capable of high-speed spins, turns, and dives in pursuit of prey. They hunt on the wing, alternating powerful wing strokes with gliding, and can hover. Primarily nocturnal and crepuscular, but sometimes hunts after sunrise and before sunset, particularly during nesting season.

Similar species: Short-eared Owl has light-colored facial disk contrasting with black eye patches surrounding the yellow eyes, only vertical streaks on the breast instead of cross-hatched vertical and horizontal streaks, and very small ear tufts that are often hidden. Great Horned Owl is much larger and more robust, with broad rather than narrow facial disk, and usually a white throat. Screech-owls are smaller and more uniform in color.

Northern Hawk Owl

(*SURNIA ULULA*)

Size: Medium. Length: 14–17.5 inches. Wingspan: 28–33 inches. Weight: 9–15 ounces.

The only member of the genus *Surnia* in the world, the Northern Hawk Owl derives its name from its hawklike behaviors, which seem somewhat emulative of falcons and of the accipiter species, such as the Sharp-shinned and Cooper's Hawks. Northern Hawk Owls frequently hunt by day, fly with direct powerful wing strokes, and can even hover. Their range spans the globe, but only at northern latitudes, making them challenging to study, which largely explains the dearth of information about their life history. Every so often, however, an irruption occurs in which Northern Hawk Owls show up far south of their normal range in winter, much to the delight of owl enthusiasts.

Northern Hawk Owl

Northern Hawk Owl

Northern Hawk Owl

Identification: Largely unmistakable, with narrow-set yellow eyes in a gray facial disk that is framed by bold black borders extending from the sides of the crown to the throat; breast/flanks/belly boldly patterned with dark bars across a white background; long, protruding tail. Relatively short, pointed wings in flight. *Fledgling:* A pale version of the adult, usually with white-tipped tail feathers even on young birds with very short tails.

Voice: Song is a rolling, drawn-out, richly toned warble, given by males during display flights; female's version is less musical, raspier, and less drawn out; also a variety of whistles and screeches.

Range of Northern Hawk Owl

Habitat: Boreal forest zone; prefers relatively open to semi-open country with ample perches for hunting, and tree thickets or forest for roosting/nesting; found in evergreen forest, mixed forest, northern hardwood forest, timbered bogs and swamps; burned forest is especially important for nesting habitat (burnt trees provide natural cavities and attract cavity-building woodpeckers).

Range: Year-round and breeding range in the boreal forest belt from Alaska to Canada's Maritime Provinces (rare); rarely breeds in northern Minnesota. During winter, especially in irruption years, a few individuals stray south of the primary range and into the Pacific Northwest, Great Lakes region, and New England.

Migration: Nonmigratory, but some birds winter south of the year-round range.

Status: Uncommon to locally common. Global population 100,000 to 500,000 and stable. IUCN Least Concern.

Diet: Small mammals, especially *Microtus* voles; also many other rodents, and Snowshoe Hares; occasionally birds such as Spruce Grouse and Canada Jays.

Nesting: Cavity nester; will use woodpecker holes and natural tree cavities, including broken-off hollowed trunks and stumps.

Clutch: Three to thirteen, 1.25-inch-long white eggs.

Behaviors: Typically hunts by perching at the tip-top of a tree or snag and watching for prey, then attacking with swift, low, powerful flight.

Similar species: Boreal Owl is much smaller and stockier, and has a short tail.

Northern Pygmy-Owl

(GLAUCIDIUM GNOMA)

Size: Small. **Length:** 6.5–7 inches. **Wingspan:** 14–15 inches. **Weight:** 2–2.5 ounces.

Despite the Northern Pygmy-Owl's deserved popularity among owl enthusiasts, this fierce, fervent-looking little owl is surprisingly enigmatic owing to a dearth of knowledge about its life history. Occupying a variety of upland habitats, and with a predilection for killing and eating other birds, Northern Pygmy-Owls are somewhat reminiscent of a distant relative, the diminutive Sharp-shinned Hawk. Both raptors are keen to take advantage of the songbird smorgasbord provided by people who maintain bird-feeding stations; "sharpies" routinely hunt such target-rich environments, and tiny Northern Pygmy-Owls provide a similar startling surprise at times by slashing in to whack songbirds that may be as big as the little marauder itself—no wonder these owls are the frequent targets of mobbing behavior by other birds.

Identification: Long tail and large, rounded head; bright yellow eyes set in poorly defined buffy face mask; brownish-gray crown speckled with white or pale tan spots, with larger light spots on wing coverts; overall grayish-brown to brown, with broad streaks on white breast/flanks. Rear view may reveal the false eyespots at the base of the neck. *Fledgling:* Like adult but with less spotting.

Voice: Distinctive call is an ongoing series of high, hollow-sounding toots, singly or in pairs, such as *toot-toot . . . toot-toot . . . toot-toot.* Toot series can extend for many minutes and may be preceded by a rapid, rich, bouncing trill. Toot songs seem to vary geographically.

Habitat: Forests, including closed-canopy forests, with openings and edges, ranging from montane conifer forests to riparian hardwood tracts to mixed woodlands; equally at home in extensive high-elevation forest and open riparian bottomlands with sufficient

Northern Pygmy-Owl

Northern Pygmy-Owl

Northern Pygmy-Owl showing false eyes on nape

Northern Pygmy-Owl

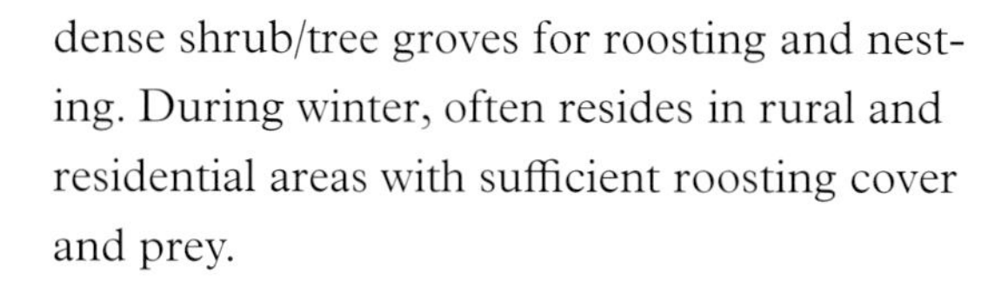

dense shrub/tree groves for roosting and nesting. During winter, often resides in rural and residential areas with sufficient roosting cover and prey.

Range: Western North America from southeast Alaska to Central America.

Migration: Probably nonmigratory, but nomadic in nonbreeding season and often moves locally/regionally to lower elevations for winter.

Status: Uncommon; population 100,000 and stable.

Range of Northern Pygmy-Owl

Diet: Small mammals, birds (including juveniles taken from nests), insects and other invertebrates, reptiles. Anecdotal reports of Pygmy-Owls attacking birds significantly larger than themselves, such as flickers and quail.

Nesting: Cavity nester; uses woodpecker holes and natural cavities.

Clutch: Two to seven, 1-inch-long glossy white eggs.

Behaviors: Lacking asymmetrical hearing, a well-defined facial disk, and silent flight, Pygmy-Owls rely on extensively on their excellent eyesight for hunting, and frequently hunt in daylight; the extent of their nocturnal hunting is unknown.

Similar species: Northern Saw-whet Owl has short tail, lighter face, and streaked (rather than spotted) crown; Boreal Owl has short tail and well-defined facial disk; Flammulated Owl has black eyes and richly toned grayish crosshatches on breast. Ferruginous Pygmy-Owls in Arizona are rich-reddish brown with fine, thin streaks rather than spots on the forehead, and usually rufous-colored tail bars instead of white.

Northern Saw-whet Owl

(*AEGOLIUS ACADICUS*)

Size: Small. **Length:** 7–8.5 inches. **Wingspan:** 18–22 inches. **Weight:** 2.6–3.5 ounces.

Common theory suggests that the name Saw-whet derives from this diminutive owl's call, but that seems a stretch when you compare the sound of a sawblade being sharpened to the Northern Saw-whet Owl's vocal repertoire. That's why the late Julio de la Torre, author and owl lover, proposed a different theory, suggesting that *saw-whet* is a Anglicized version of the French word for a small owl—*chouette*, the name used by French-speaking Nova Scotians,

Northern Saw-whet Owl

Northern Saw-whet Owl

Northern Saw-whet Owl fledglings

who pronounce the *c* as an *s*, to refer to the Saw-whet Owl, hence "sou-ette."

Identification: Tiny, about the size of an American Robin; creamy-white breast/flanks patterned with broad, pale to dark, rusty-brown streaks, sometimes gray; dorsal surface typically rusty-gray, sometimes more rusty-brown or slate gray, with prominent white spots along the wings and scapulars; head appears oversized, with bright yellow eyes set in a pale face mask with modestly defined borders of feathery dark/light streaks radiating outward; dark crown is finely streaked and bill is black. *Fledgling:* Brown facial disk with broad white V-shape eyebrows and cinnamon-colored breast/belly/flanks.

Voice: Common song is a lengthy series of evenly spaced high-pitched toots, about two per second; other sounds include

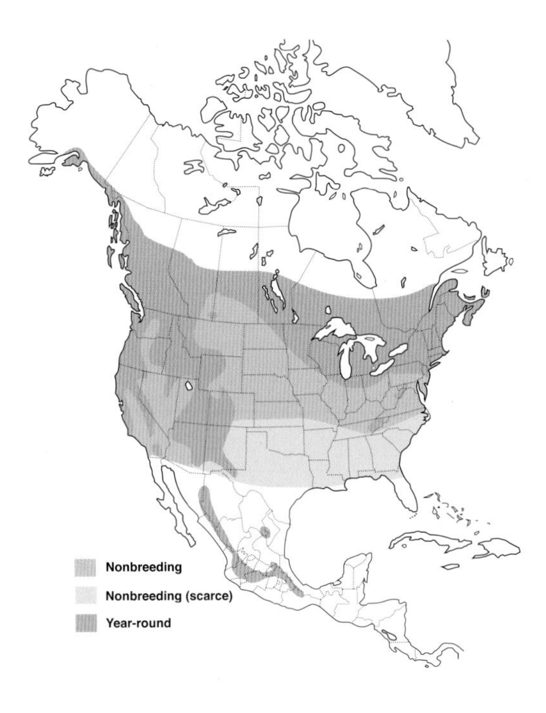

Range of Northern Saw-whet Owl

exotic-sounding high-pitched yelps, eerie wails, and high-pitched screeches.

Habitat: Coniferous and mixed forest.

Range: Breeding and year-round range from Alaska to southern Mexico and from the West Coast to New England, but absent as a breeder in the South and the interior of the United States and southcentral Canada; during winter, range extends through most the United States.

Migration: Partially migratory, with some populations and individuals withdrawing southward and some to lower elevations.

Status: Common. Population 2 million. IUCN Least Concern.

Diet: Primarily small mammals, such as mice, voles, and shrews; occasionally small birds and invertebrates.

Nesting: Cavity nester; uses woodpecker holes and natural cavities.

Clutch: Five to seven, 1.2-inch-long white eggs.

Behaviors: This widespread little nocturnal owl is perhaps the most frequently banded owl species in North America, according to The Owl Foundation (www.theowlfoundation.ca): during migration, Saw-whets fly low and funnel through specific narrow corridors, where they are easily intercepted by licensed bird banders; data gathered from banded and recaptured Saw-whet Owls is helping us better understand their migration routes and timing, which in turn can lead to enlightened conservation decisions.

Similar species: Larger Boreal Owl is typically grayer overall and has a pale-colored bill, well-defined facial disk, and white spots rather than streaks on the crown.

Spotted Owl

(*STRIX OCCIDENTALIS*)

Size: Medium. **Length:** 17–19 inches. **Wingspan:** 43–46 inches. **Weight:** 16–30 ounces.

The Spotted Owl is in trouble, especially in the Pacific Northwest, which is home to one of the three subspecies, the Northern Spotted Owl. In the 1980s, these birds became iconic symbols—beloved or hated, depending upon one's sociopolitical leanings—of the struggle to curtail industrial logging of publicly owned old-growth forest in the Northwest. Indeed, these oft-tame dark-brown owls are symbolic denizens of the vastly shrunken old-growth Douglas-fir/mixed-conifer forests that once carpeted the mountains of British Columbia, Washington, Oregon, and Northern California. And even as efforts to preserve some tracts of this unique ecosystem succeeded, an ominous new threat was already emerging: larger and closely related Barred Owls expanded their range, invading the forests of the Northwest, where they drive off, kill, and hybridize with Spotted Owls.

Northern Spotted Owl

California Spotted Owl

Identification: Overall mottled dark brown, with rows of horizontal oblong white spots on the breast and flanks, and small white spots on the brown mantle and crown; all-dark eyes, no ear tufts; grayish-brown facial disk and usually prominent white X between the eyes. Northern Spotted Owl (*S. o. caurina*) is the darkest of three subspecies; California Spotted Owl (*S. o. occidentalis*) is intermediate, and Mexican Spotted Owl (*S. o. lucida*) is palest and

Mexican Spotted Owl

Northern Spotted Owl

smallest. *Fledgling:* Pale gray ventral side and head; chocolate brown mantle and dorsal wing surfaces with whitish spots; developing facial disk looks like two dark eyebrows.

Voice: Characteristic song is a series of four rich, resonant hoots, the two middle notes stacked together, *hooo . . . hoo-hoo . . . hoo*, and often preceded and/or proceeded by additional hoot notes; other calls include high-pitched wails, a rising whiny whistle, and sharp high-pitched barks.

Habitat: *Northern Spotted Owl:* Mature and old-growth conifer forest. *California Spotted Owl:* Elevation dependent, from oak and other hardwood forest at low elevations to mixed forest to mature conifer forest at high elevation. *Mexican Spotted Owl:* Mature conifer and pine/oak forests, and canyon riparian forest.

Range: *Northern Spotted Owl:* Cascade Range, Siskiyou Mountains, and Coast Ranges of the Pacific Northwest from southern British Columbia south to Northern California. *California Spotted Owl:* Southern Cascades/ northern Sierra Nevada southward through noncontiguous mountains of Central and Southern California. *Mexican Spotted Owl:* Interior West, including southern Utah, central and southern Colorado, Arizona, New Mexico, west Texas, and south to central Mexico.

Migration: Nonmigratory.

Status: *Northern Spotted Owl:* Rare and in rapid decline, with population decreasing so quickly that the population is difficult to estimate, but probably comprises 4000 or fewer mature birds. *California Spotted Owl:* Rare and probably declining; population unknown. *Mexican Spotted Owl:* Rare and probably declining, population unknown. Total Spotted Owl population (North America including Mexico) estimated at 15,000 and declining. IUCN Near Threatened, but extremely and increasingly rare and likely soon extirpated from British Columbia, and rapidly declining in Washington and Oregon.

Diet: Primarily small mammals, but diet composition depends on location; for example, Northern Spotted Owls are significantly dependent on flying squirrels, but farther south, wood rats are more important; other important prey includes voles, mice, pocket gophers, birds, rabbits, squirrels, and even bats.

Nesting: Nests in natural tree cavities, especially broken trunk hollows in large dead trees, as well as platforms such as stick nests built by other birds, and mistletoe brooms; sometimes cliff hollows and platforms.

Clutch: One to four, 2-inch-long white or light gray eggs.

Northern Spotted Owl fledgling

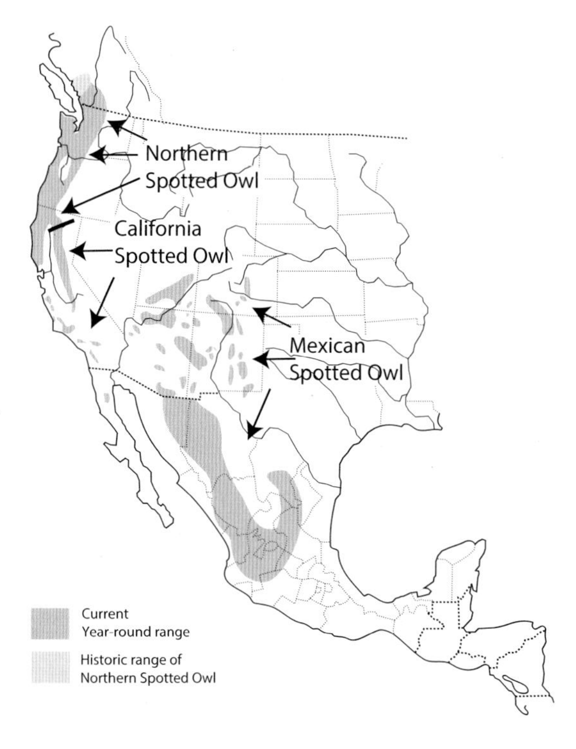

Range of Spotted Owl

Behaviors: A pounce-and-perch predator, the Spotted Owl is an agile but deliberate flier that sits on subcanopy branches to watch and listen for prey before striking from above by gliding silently in for the attack.

Similar species: See Barred Owl.

Short-eared Owl

(ASIO FLAMMEUS)

Size: Medium. **Length:** 13.5–17 inches. **Wingspan:** 36–44 inches. **Weight:** 8–16 ounces.

One of the world's most widespread owls, the Short-eared Owl is a denizen of open spaces, where it hunts on the wing and nests on the ground. The males perform a dazzling courtship flight, circling to ascend high into the air—up to 1000 feet sometimes—then hooting excitedly as they fly, before plunging with folding wings and making audible wing claps. They may repeat the display several times, with a female presumably watching from the ground. These display flights hint at their aerial agility: Short-eared Owls quarter just above ground level in search of prey and can hit the brakes and turn on a dime or plunge suddenly to snatch a rodent or bird. Worldwide there are 11 subspecies of Short-eared Owls; North America is home to the Northern Short-eared Owl, and the Hawaiian Islands boast their own unique subspecies, called the *Pueo* in the native language.

Short-eared Owl

Identification: About the size of a crow, Short-eared Owls are overall heavily streaked in buff, tan, and brown on both the ventral and dorsal surfaces, with the breast/belly marked with vertical streaks (the Long-eared Owl has crosshatched streaks); their tiny ear tufts are often invisible. In the northern race, found in most of North America, the yellow eyes contrast with striking black eye patches, as if the bird is wearing heavy mascara; the eyes and eye patches contrast with a pale facial disk often handsomely framed with lighter feathers. Face color is variable, however. The bouncy flight pattern, often described as resembling the flight style of a butterfly or moth, is distinctive. *Fledgling:* Like adult but paler breast and belly, and often with dark shades around the eyes and bill.

Short-eared Owl

Voice: A short, scratchy, bark-like screech is common, along with a lengthened but similar scream, which can ascend or descend in pitch: *keeeer-IP* or an almost catlike *keeer-ow*. Males hoot, usually in a long, surprisingly deep series, often quickening, *hoot-hoot-hoot-hoo-hoo-hoo-hoo-hoo*.

Habitat: Short-eared Owls hunt over open habitats, including plains and grasslands, salt- and freshwater marshes and estuaries, tundra, farmlands, sagebrush- or other shrub-dominated steppe; breeding, migration, and wintering habitats are similar.

Range: Circumpolar; Short-eared Owls are widely distributed around the globe, especially in the Americas and Eurasia. In North America, they occur from northern Mexico to

Short-eared Owl

Alaska, and from coast to coast. They winter in the southern half of that range and summer in the northern half, with a substantial swath of overlap where they can occur year-round. They inhabit many Caribbean islands as well as Hawaii.

Migration: Withdraws southward from Alaska and Canada for winter, and winters across much of the United States, irregularly, as far south as Arizona, New Mexico, and northern Mexico. Typically arrives in

Range of Short-eared Owl

northernmost breeding range by the end of April and departs in November.

Status: Uncommon to locally common, and widespread. Global population 1.2 to 2 million mature birds and declining; IUCN Least Concern.

Diet: Mostly small mammals, especially voles, but also small to medium-size birds.

Nesting: Nests on the ground, in the open or within concealing vegetation, and builds its own nest by hollowing out a shallow scrape and lining it with soft vegetation and feathers.

Clutch: Five or six (sometimes up to more than a dozen), 1.5-inch-long white to slightly off-white eggs.

Behaviors: Like Barn Owls, Short-eared Owls generally hunt in flight, coursing low over open areas much in the fashion of the Northern Harrier, a wide-ranging hawk of about the same size (see Similar species). Their bouncy, butterfly-like flight pattern is distinctive, and they often hover before plunging on prey. They are well-known for hunting during the day, especially early morning and evening, but also midday; however, nocturnal and crepuscular hunting is more routine but less observable. During the nonbreeding season, Short-eared Owls sometimes roost communally in favorite trees or ground roosts near prime hunting areas, and sometimes even roost alongside Long-eared Owls. Ground-roosting and especially nesting Short-eared Owls often flush only when closely approached.

Similar species: The Long-eared Owl has a tawny facial disk (whitish facial disk with black eye patches in the Short-eared Owls found in most of North America) and its long, pronounced ear tufts are frequently visible; Northern Harriers have long tails and large, distinctive white rump patches (female and juvenile Northern Harriers are brownish overall, males are grayish overall); the Barn Owl is white to nearly white below, has a heart-shaped white facial disk, and has black eyes (yellow in the Short-eared Owl).

Snowy Owl

(*BUBO SCANDIACUS*)

Size: Large. Length: 21–28 inches. Wingspan: 50–57 inches. Weight: 3.5–6.5 pounds.

Strikingly beautiful in their all-white to dalmatian-patterned plumage, and among the largest and heaviest owls in the world, Snowy Owls are iconic denizens of the far north, where they nest on open, windswept tundra. But during the nonbreeding season, Snowy Owls significantly expand their range southward, which gives many more people opportunities to observe these magnificent owls in the wild. Further, in some winters, Snowy Owls are irruptive, meaning large numbers of them wander south into southern Canada and the northern tier of the Unites States, and a few show up deep into the southern expanses of the United States, always to great fanfare. Irruption was once thought to be a population-wide response to a shortage of prey, particularly lemmings, but researchers have since discovered that an abundance of food triggers Snowy Owl irruptions. When prey abounds during the arctic nesting season, nesting success improves and brood counts increase, meaning lots of young-of-the-year Snowy Owls capable of heading south for the winter. Modest irruptions occur every four or five years; mega-irruptions are rare but occur perhaps every two or three decades.

Snowy Owl, female

Snowy Owl, male

Snowy Owl, juvenile

Snowy Owl, female

Identification: Large and predominantly white to partially white, with variable black or brown markings, and an all-white face; females and juveniles typically have substantial dark spots and chevrons, while adult males are often starkly white with minimal black spotting. Smooth, rounded head that only rarely shows small ear tufts. White face lacks a darker frame, such as in the Barn Owl; yellow eyes; bill well covered with white feathers; legs and toes covered in soft white feathers. *Fledgling:* Overall gray, with white/black barred wing feathers; gray replaced by white with dark spots/patches as the bird ages.

Voice: On breeding grounds especially, males call with a series of two (sometimes more) low hoots, and grating hoots. Females rarely hoot but emit harsh, high-pitched screams and shrieks. Other calls include low chuckles, raspy barks, and even defensive bill clicking. During the nonbreeding season, they are mostly silent.

Habitat: Nests on arctic tundra, but disperses to a wider range of open habitats during nonbreeding season from mid-autumn through March or April; during nonbreeding season, Snowy Owls expand their habitat to include seashores and lakeshores, dunes, croplands, expansive fields, prairie and steppe, airports, and other places with minimal cover and broad vistas. They are famous for wintering in both rural and urban areas far to the south of their breeding range and will sometimes perch on rooftops and other human-made structures.

Range: Circumpolar in the arctic, with North American breeding range in the High Arctic and year-round range (including nesting range) in the adjacent Canadian Low Arctic and northwesterly across the northern and northeast edges of Alaska; North American nonbreeding range is southward through Canada (coast to coast) and into the northernmost United States. During irruption years, wintering Snowy Owls routinely show up farther south.

Migration: Significant numbers withdraw southward for winter; some remain in the breeding range during winter, and some even migrate northward out onto the sea ice, the land of 24-hour nights during the long, cold, arctic winter.

Status: Rare to locally common at low densities. Global population 14,000 to 28,000 mature birds and declining; IUCN Vulnerable, with numbers declining because of human development and climate change.

Diet: On their arctic nesting grounds, Snowy Owls feed extensively on lemmings, and also eat other rodents, arctic hares, and ptarmigan and other birds, and being opportunistic, just about anything they can kill. Iceland, however, which has no lemmings, has a very small resident population of Snowy Owls, and these owls feed primarily on birds, especially Rock Ptarmigan and various shorebirds. Snowy Owls are adept at catching birds in flight, as well as on the ground or water. During the nonbreeding season, Snowy Owls throughout the species' range feed on a wide array of prey. Some routinely kill waterbirds on open water, especially at night when seabirds and ducks sleep: researchers with Project SNOWstorm collect telemetry data from owls fitted with transmitters and found that some of the tagged owls, on their wintering grounds, not only spend lots of time hunting over open water but also use channel markers and other such structures as perches.

Range of Snowy Owl

Nesting: Ground nesters, Snowy Owls generally choose a slight rise or hummock on the tundra—a place exposed to the wind, which keeps biting insects at bay and prevents snow from accumulating as easily, and which is high enough to remain dry while the surrounding slightly lower areas can be boggy. Moreover, the high mounds provide a vantage point from which the owls can see danger (such as various nest predators). These ubiquitous mounds are called high-centered polygons, or *pingaluks*. The female builds a modest depression in the ground atop the higher mounds, sometimes through remaining snow and ice, and may choose the same nest site for successive years; she lays from three to eight eggs, which hatch in approximately 32 days, but can lay a dozen or more in years of especially high prey abundance.

Clutch: Three to eight (sometimes up to 12), 2.2-inch-long white eggs.

Behaviors: Snowy Owls spend a lot of time perching on the ground or on any object that provides them a good view of the surrounding area; they hunt day and night, with nocturnal hunting commonplace during the long nights of late autumn through mid-spring, the time period when most people see Snowy Owls (in locales well south of the arctic); wintering Snowy Owls take up temporary residence in areas with ample open countryside or open water, and frequently hunt at night. Conversely, because the sun remains visible all or most of the day in the arctic during the Snowy Owl nesting season, the birds must hunt in broad daylight and during the crepuscular hours of night when the sun is low to the horizon. Moreover, they must bring a constant supply of lemmings or other prey to the owlets until the young are fledged.

Similar species: Although the two species are dissimilar, Barn Owls are sometimes misidentified as Snowy Owls by casual observers because Barn Owls are predominantly white on their undersides; when Barn Owls are caught briefly in automobile headlights or other bright lights, witnesses often report a pure-white owl. However, Snowy Owls are much larger and more robust than Barn Owls.

Western Screech-Owl

(MEGASCOPS KENNICOTTII)

Size: Small. **Length:** 8–9 inches. **Wingspan:** 20–22 inches. **Weight:** 4–10 ounces.

The counterpart to the widespread Eastern Screech-Owl, the Western Screech-Owl is a fierce little predator, but its diminutive size means it also must always be on the lookout lest it become prey for something bigger, such as a Great Horned Owl. One

Western Screech-Owl, gray-brown Pacific Northwest form; ear tufts lowered

Western Screech-Owl, interior form, ear tufts raised

defense mechanism is excellent camouflage—Screech-Owls often perch at the entrance to roost holes or notches or crevices in trees, where their mottled gray plumage blends seamlessly with bark patterns.

Identification: Light gray-and-white to light buff-and-white breast and flanks marked handsomely with black crosshatched streaks; overall shade ranges from rich gray/white/black to tawnier instead of gray (gray-brown morph), especially in the Pacific Northwest. Bright yellow eyes set in gray facial disk that is framed with black to varying degrees; dark gray-black bill; fairly small but prominent ear tufts can be lowered; stocky appearance. *Fledgling:* Similar to adult, but gray/white pattern on the breast/flanks composed mainly of horizontal bars much like the pattern of a Plymouth Rock chicken.

Western Screech-Owl, Southwest form, ear tufts lowered

Western Screech-Owl fledglings

Range of Western Screech-Owl

Voice: Commonly called a "bouncing ball" song, the Western Screech-Owl's distinctive call is a series of 8 to 12 or more toot notes increasingly close together, *toot . . . toot . . . toot . . . toot-toot-tootootootootoo*; also a variety of shrieks, trills, and barks.

Habitat: Deciduous and mixed forest, including riparian zones; extensive tree or shrub thickets in otherwise open country; farmlands; and even urban areas with suitable tree cover.

Range: Western North America from southwest Alaska to central Mexico.

Migration: Nonmigratory.

Status: Uncommon to locally common; population unknown. IUCN Least Concern.

Diet: Insects and other invertebrates, small mammals, small birds, crayfish, fish, amphibians, reptiles.

Nesting: Cavity nester that uses woodpecker holes and natural cavities and will use artificial nest boxes.

Clutch: Two to four, 1.3-inch-long whitish eggs.

Behaviors: Mostly a perch-and-pounce predator, but can hawk insects (and bats) on the wing.

Similar species: The Western Screech-Owl's range only minimally overlaps with that of the very similar Eastern Screech-Owl, mostly in relatively small areas of Texas and Colorado; the two species have different calls (see descriptions). Hybrids are known to occur in the range-overlap areas. The smaller Whiskered Screech-Owl is very similar in appearance to the Western Screech-Owl, but its calls are different, and its northernmost range extends only to southeast Arizona and barely into northwestern New Mexico. The Flammulated Owl is smaller and has black eyes.

Whiskered Screech-Owl

(MEGASCOPS TRICHOPSIS)

Size: Small. Length: 6.5–8 inches. Wingspan: 16–20 inches. Weight: 3–4 ounces.

The enigmatic little Whiskered Screech-Owl haunts the montane nights of Central America, Mexico, and the isolated highlands of the deep Southwest. This is perhaps the least-studied owl found in the United States, owing to its limited range north of the Mexican border and its preference for high-elevation woodlands; however, most studies on this species have occurred in Arizona, rather than in their lengthy range through the mountains stretching down to Nicaragua. Primarily an insect eater, the Whiskered Screech-Owl has adapted well to suburban environments within its upland canyon habitat, where streetlights and other such illumination attract its favorite prey. This owl is named for the whiskerlike feather extensions on its face, but they are difficult to see.

Identification: The smallest of the three screech-owls that inhabit the United States, the Whiskered Screech-Owl is overall like the Western Screech-Owl, and the two species overlap in range in southern Arizona and southwestern New Mexico (and south of the United States). The Whiskered Screech-Owl has bolder black-on-silvery gray plumage on the breast and often with variable buffy tones on the face, throat, and/or upper breast. Its prominent ear tufts are often concealed; the bright yellow eyes are typically slightly more golden or yellow-orange than yellow, but such characteristics are extremely difficult to judge in the field. Note the olive-colored bill (blackish in the Western Screech-Owl). *Fledgling:* Similar

Whiskered Screech-Owl

Whiskered Screech-Owl

to other Screech-Owl species, with gray to rust-gray horizontal bars patterning the breast and flanks.

Voice: Often described as reminiscent of Morse code, the typical call is a series of speedy shrill hoots, and the male's song is a more even, slowly delivered series of hoots. Note that the common call of the Western Screech-Owl is a high-pitched "bouncing ball" hooting, accelerating at the end of the series. Like most owls, Whiskered Screech-Owls have a litany of other vocalizations.

Habitat: High-elevation woodlands, especially in riparian canyons in Arizona's southern mountain ranges, including pine/oak/sycamore forest; nesting habitat is generally between 5000 and 8000 feet or higher in elevation.

Range: The Whiskered Screech-Owl reaches its northernmost range in the sky-island mountains of southeastern Arizona and the southwestern corner of New Mexico; from there, its range extends south through the mountains of Mexico and to central Nicaragua. Within the United States, this small owl is common only in the southern Arizona mountains. Most sightings occur in the Chiricahua, Huachuca, Patagonia, and Santa Rita Mountains, but the species occurs as far north as the Santa Catalina Mountains northeast of Tucson and as far west as the Tumacacori Highlands. Most New Mexico sightings come from the Peloncillo Mountains.

Migration: Nonmigratory.

Status: Locally common at high elevation in the mountains of southern Arizona. Population in the U.S. estimated at fewer than 500 and declining; global population unknown. IUCN Least Concern.

Diet: Primarily an insect eater (especially beetles and caterpillars), Whiskered Screech-Owls will also eat other invertebrates, small reptiles, rodents, and even the occasional bird and bat.

Nesting: Nests in cavities, including woodpecker holes and natural cavities in tree trunks; data from Arizona found that Arizona Sycamore (*Platanus wrightii*) is a common nesting habitat, but that data does not mean the owls are selecting for tree species, but instead may testify to the availability of cavities in Arizona Sycamore and/or relative abundance of this tree species near water, where the owls prefer to nest. For several years starting in 1979, researchers built and monitored nest boxes intended for Elegant Trogons (a stunningly colorful bird that shares the same mountain-canyon habitat in southeast Arizona); no trogons used the nest boxes, and only one box was utilized by Whiskered Screech-Owls and one by Flammulated Owls, perhaps indicating that both natural and woodpecker-excavated tree cavities are sufficiently abundant within the range of these species.

Clutch: Two to four, 1.1-inch-long whitish eggs.

Behaviors: Strictly nocturnal; during the day, tends to perch within the protective shelter of dense foliage or branches and/or close to camouflaging tree trunks.

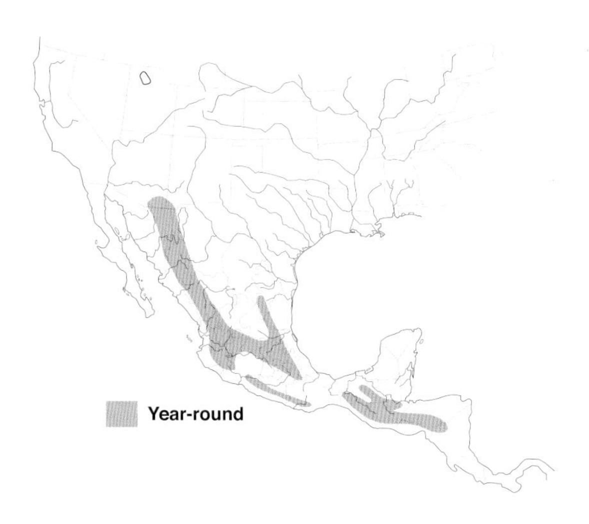

Range of Whiskered Screech-Owl

Similar species: Western Screech-Owls are nearly identical to, though larger than, Whiskered Screech-Owls; voice is the best way to distinguish the two species (see descriptions). Whiskered Screech Owl has slightly bolder dark markings on the breast and often has soft rust tones on the throat; both species usually have light-tipped bills, but the basal portion of the Whiskered Screech-Owl's bill is dusky olive rather than black, as in the Western Screech-Owl. Still, differentiation in the field without hearing diagnostic calls is difficult without excellent views. The Flammulated Owl has black rather than yellow eyes; Elf Owl is smaller, less boldly patterned, and primarily occurs at lower elevations than the Whiskered Screech-Owl.

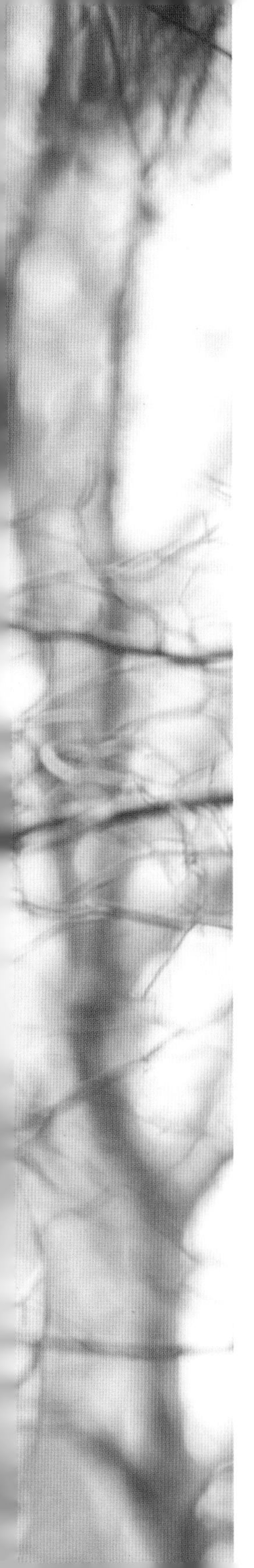

3

The Art of Owling

Low-growing sagebrush, the kind battered by wind and icy winters, huddles on the slopes plunging into a shallow canyon winding through a remote high desert in northern Nevada. Thrice over its 2-mile course, this rocky cleft narrows into true box canyons in miniature, like small-scale replicas of famous western landscapes.

On an early summer day many years ago, my wanderlust convoked with my love of Great Basin deserts and I decided to see what natural secrets this nondescript gorge might hide from the public speeding by just a few hundred yards distant on a little-used backcountry highway. A detailed topographical map provided a bird's-eye view of the basic layout, so I laced up well-worn hiking boots, donned a heavy shirt against the chill of a desert dawn, and set out down the bottom of the canyon. Starting at its shallow upstream end, where a dry streambed—probably perennial 13,000 or more years ago when Native Americans first roamed these lands—slanted down from a wide valley, I soon found

Barred Owl

geological treasures aplenty, a rockhound's dreamscape if only I were a rockhound. And wildflowers—a trove of intriguing wildflowers, some of which I'd seen nowhere else.

The ancient streambed, studded with sun-baked, water-polished rocks, occasionally gave way to sandy expanses abounding in animal tracks: mule deer, coyotes, bobcats, and pronghorns had passed this way; bighorn sheep, sage-grouse, and Chukar Partridge, too. Mine were the only boot prints.

Half a mile down the gradually deepening canyon, a long straightaway sank into a sharp S-curve where dramatic hoodoos jutted skyward from crumbling stone parapets, testament to the erosive powers of water and wind over eons. Waterworn slabs, where the creek once formed cascades, served as stairsteps down into the chasm beneath the hoodoos, and when the streambed's left-hand S-curve reversed itself, I entered a wonderland, a classic box canyon, with vertical walls guarding both sides. Though a mere 50 yards long, this walled-in section of the canyon hid a water hole—a sunken pocket tucked underneath the south wall, hidden from the sun, and pocked with animal tracks.

Taking in the scene and understanding the value of a natural water hole in this desert environment, I was startled when a bird took flight from the canyon wall just a few feet from where I stood. An owl. I had accidentally interrupted its daytime slumber and it sped silently away, angling up and over the top of the rim. Though I'd only gotten a fleeting view of the bird, I was reasonably sure it was a Western Screech-Owl; but its presence in that environment seemed a non sequitur. This was my inaugural foray into that hidden canyon, but I'd been exploring the surrounding region for two decades and I couldn't think of a single copse of trees, riparian or otherwise, within miles of where that owl had roosted. Western Screech-Owls like trees; they like groves of trees, where they can roost during the day, allowing their cryptic plumage to blend with bark and branches.

This intriguing encounter—a Western Screech-Owl where I had assumed one shouldn't be—triggered my mild obsessive-compulsive disorder, and the next year at about the same time in June, I retraced my steps down to the hidden water hole very early in the morning. I had to be sure. Perhaps I'd seen a Long-eared Owl and simply misjudged the size. This time, camera at the ready, I snuck carefully around the bend in the old streambed, but I wasn't stealthy enough, and once again a little owl dashed away, down the box, out the other end, and up over the rocks. I missed the chance at photographs, but confirmed by a better view, brief though it was, that the bird was indeed a Western Screech-Owl. Examining the rock face, I found white excrement (aka whitewash) in abundance, dribbled down the rocks in several places, and built up on the ground below. Possibly it was owl whitewash, though I couldn't rule out other raptors, such as Prairie Falcons.

A year later I returned, but decided to remain at a significant distance from the owl's roost. I approached from above, hiking along the north rim of the canyon until I neared the boxed-in section, then I headed northward about 200 yards to a promontory from which the upper third of the canyon's south wall was plainly visible. I sat on the ground and with binoculars scanned what I could see of the wall, but found no owl. I wondered what had become of the seemingly out-of-place screech-owl, though I was aware that small owls often live only a few years, with a potential lifespan of perhaps eight to 10 years. Nonetheless, I had enjoyed those three days, each about one year apart, of "owling" in the canyon.

Owling usually means searching for owls by sight or sound. Owlers comprise professional researchers, hobbyist bird-watchers, and people who are simply curious about owls. Owling can take various forms, from targeting a rare or seldom-seen species to field trips planned with the aim of finding several species. Rarely does owling denote the singular pursuit to photograph one individual owl, but my fascination with that Western Screech-Owl in the Nevada desert is not unique—others have been likewise inspired by individual owls.

But the experience was revealing, especially regarding owls and their adaptability. Despite a dearth of typical Western Screech-Owl habitat, this individual (and perhaps his or her kin) had found a desert environment to its liking, substituting mottled grayish-brown cliff walls for the more-typical mottled grayish-brown tree trunks and limbs for its camouflaged day roost. The cliffs, hoodoos, and rock bluffs in that area are pocked with holes of all kinds and sizes—roundish divots of varying depths, channels, slots, natural tunnels, overhangs. I surmise some of these cavities would serve admirably as nest sites for Western Screech-Owls.

The year after I failed to find my little screech-owl in the box canyon, I decided to place a trail camera (aka game camera) at the water hole at the base of the south cliff. The idea wasn't inspired by the owl; rather, I was simply curious about what kinds of creatures might use the water hole during the hot summer. On a warm day in late April, I snuck quietly toward the box, approaching from above so as not to disturb any roosting birds. I found no owl (but enjoyed watching the seemingly ever-present Rock Wrens dashing about on the rocks and singing ebulliently), so I climbed down and placed the camera at the water hole.

The location was beyond the reach of cell phone coverage, so I could not check the cam remotely, a fact that only added to the intrigue as the weeks passed until late July when I had the opportunity to retrieve the camera. I was amazed to find that I had captured more than 4000 images: Chukar Partridge, Common Ravens, Red-tailed Hawks, Turkey Vultures, songbirds, wood rats, cottontails, bobcats, a badger, mule deer, bighorn sheep, bats, coyotes, and much to my surprise and joy, a Great Horned Owl with its fledgling. In one sequence of images, the adult owl seems to be showing the juvenile how to drink from and bathe in the water.

The presence of a Great Horned Owl would make that canyon a perilous place for any smaller species of owl, so I wondered if I'd inadvertently discovered why the screech-owl had been a no-show for the past two years. In any event, finding that Western Screech-Owl in an unlikely place, and capturing trail-cam photos of the Great Horned Owl and its fledgling, gave me a better understanding of these birds. Subsequently, within a 10-mile radius of the canyon, I found nesting Long-eared Owls and nesting Short-eared Owls. I had always enjoyed birds and bird-watching, and I'd never forgotten that snowy night in Idaho when Dad shone his flashlight on the Great Horned Owl perched on the wooden power pole; a few years hence I would become passionately fascinated by hummingbirds, but in a remote part of Nevada, I became an owler.

The Ethics of Owling

Owls attract attention. They elicit awe and interest from all sorts of people, from dedicated bird-watchers to the casually curious among us; they are popular targets for photographers. We relish opportunities to observe owls during the day because these birds can be so cryptic, so difficult to find, and so challenging to observe when they are active after dark. An owl roosting in a tree or a cliff or an old barn is usually trying very diligently not to be seen, and thanks to their camouflaging plumage and secretive nature, owls usually evade detection by human eyes.

So how are we owlers to find the objects of our affection? The answer to that question begins with what *not* to do. Find any of the less-often-seen species of owl, announce the location by any form of social media or online platform, and watch what happens: almost immediately the site where the bird has been found will attract other people, often lots of other people. Some will behave appropriately, keeping the owl's best interest in mind; others will violate that ethic. One winter day many years ago, a Snowy Owl arrived in a small Northwest town and chose a homeowner's rooftop as its preferred day roost. A big white owl sitting on a roof can hardly go unnoticed,

When this Snowy Owl (perched atop the power pole) began attracting admirers in southern Canada, the photographer kept his distance, not wanting to unduly disturb the bird, but many other onlookers got too close. Even if an owl does not flush and fly away, too many disturbances may interfere with its ability to find food and avoid predators.

and soon the quiet neighborhood was swarming with birders, many wielding cameras, and most clamoring for an ever-closer view. No doubt most of the curious onlookers tried to be unobtrusive, but others seemed unconcerned about the privacy of the residents or the well-being of the owl, which finally disappeared, hopefully to find a less stressful winter retreat.

Owls are easily disturbed and easily stressed into altering their behavior at great peril to themselves. All owls must navigate the razor's edge balance of energy exchange: they must find enough food, for themselves and their

This Western Screech-Owl uses its concealment posture, stretching out to better blend with tree branches.

young, to offset the energy expenditure of hunting for food and other activities. Disturbing an owl risks tipping that balance in the wrong direction. Moreover, North America's smaller owls—meaning just about any species other than Great Horned Owls and Snowy Owls—are subject to predation by big owls and other raptors, and many different predators are only too happy to raid an owl nest to dine on eggs or hatchlings. When we disturb an owl, causing it to alter its behavior, we may inadvertently make the bird or its nest more vulnerable to predation.

Luckily, owls are reasonably adept at telling us when we've crossed the line—when we are too close, too loud, or too obnoxious for their liking. Some species can indicate their anxiety by body language: if an owl slims down by stretching its body lengthwise, it is attempting to better blend in with surrounding limbs and tree trunks to evade detection. Transitioning from relaxed and asleep to slimmed down is a sure sign an owl is disturbed, and this commonly occurs when people approach a roosted owl too closely. Slimming down is typically accompanied by raised, straight ear tufts (in species such as screech-owls and Long-eared Owls), and eyelids closed to slits—this combination of behaviors is called a concealment posture.

A disturbed, stressed owl may begin bobbing its head, and may gently pump its wings and appear ready to launch into flight—and will soon do just that if observers don't quickly back away. Some owls, particularly larger species, sometimes fan their feathers so they look much larger in response to a threat. This behavior seems to be most common with an injured or otherwise incapacitated owl, particularly if found on the ground, but is also a defensive ploy used by healthy adult and juvenile owls. In fact, years ago, while slowly driving a desert road at night, I rounded a bend and my headlights illuminated a Great Horned Owl on the ground; the bird immediately puffed up its feathers and spread its wings before flying off with the kangaroo rat it had killed. The sudden appearance of two beaming headlights had understandably startled the owl, which was happily unwilling to give up its catch.

A roosting owl may also simply flush and fly away when people get too close. Smaller owls put themselves at significant risk when they leave the concealment of a roost during daylight hours, when bird-killing raptors hunt: Cooper's Hawks, Northern Goshawks, various falcons, and even ubiquitous Red-tailed Hawks are all serious threats to any small owl that ventures out during daylight. And owls of all sizes risk death or injury by being struck by vehicles, a threat that is magnified when roads are busy during the day. Collisions with vehicles kill and injure countless owls every year.

As owl fans, we need to temper our eagerness and monitor our own behavior to avoid putting these birds at risk by our activities. The overriding ethical rule is to do no harm, and that ethos encapsulates rules we can all follow to make sure owls (and other birds) can go about their lives undisturbed. Guidelines for observing owls without disturbing them are aptly codified by various organizations, including the Minnesota-based International Owl Center, whose precepts are included here.

In addition to these guidelines, I would redouble the warning against hotspotting, the act of announcing an owl's location on social media sites or online forums. We all want to see owls, and I'll take any tip I can get, but outing the location of roosting or nesting owls is almost certain to draw a crowd of onlookers, and they won't all abide by birding ethics.

This Barred Owl, accustomed to humans, was perched alongside a well-used trail, so the photographer took the opportunity to snap a few photos and then moved on.

In addition, plan your owling adventures for the nonbreeding season between mid-autumn and mid- to late winter, before owls are incubating eggs and caring for young. The less disturbance they are subjected to, the greater the parents' chances of successfully rearing a new generation of owls.

Photographers also need a code of ethics, and wildlife photography is rampantly popular, thanks to digital technology that has made capturing great images easier than ever before. In the days of film photography, an assembled crowd at a popular birding site would include a few photographers with big telephoto lenses; nowadays almost every birder in such a gathering will be armed with a camera. But the bird's well-being must always trump the desire to snap captivating images. Photographers need to remain at the same respectful distance from an owl as nonphotographers, and never manipulate the scene by trying to make the bird turn, move, or fly away.

Owls and High Tech

Remote Cameras

From my experience at the hidden canyon-bound water hole in Nevada, I realized (belatedly by comparison to many other owl fans) that an unobtrusive remote trail camera is another great way to watch and learn about owls. These cameras allow you to peer into the night, to catch a glimpse of the owl's nocturnal world without disturbing the birds. Operating trail cams, however, carries with it a set of ethics, and in some states requires adherence to specific laws. Be sure to check with your state's wildlife management agency for regulations.

One major problem that has led to new laws governing the use of trail cams is that hunters can give themselves an unfair and unsporting advantage with remote cameras, even more so with modern trail cams that provide live feed to a cell phone, tablet, or computer. And without a remote feed, cam operators might be tempted to visit their camera to retrieve images too often, thereby disrupting wildlife. These problems are especially acute in the arid American West, where trail cams have become so popular with hunters that at times well-known water holes are practically ringed with cameras. That's why western states have increasingly adopted stricter regulations. Nevada, for example, now bans trail cams on public land from August 1 to December 31, a period that coincides with big game seasons; the ban begins July 1 for cameras capable of transmitting images, video, or locations of wildlife.

Trail-cam ethics cover two categories: the expectation of privacy or lack thereof for people on public lands, and the necessity to cause no disruption of wildlife. Again, with rapidly evolving camera technology and the burgeoning popularity of trail cams, states are beginning to formulate regulations governing their use. I don't want to disturb anyone's privacy on public lands, and I don't like the idea of someone's wildlife camera inadvertently spying on me. So, if I place a trail cam on public land, it's always in remote places and never along established trails or roads. The same principles

(continued on page 142)

How to Behave Around Owls

BY THE INTERNATIONAL OWL CENTER

Give them space. The majority of owls don't like being around people, although some in urban areas become habituated to humans and tolerate fairly close approach. Use cameras with long lenses. Don't try to get a good photo with a cell phone, since you will very likely have to get close enough to scare the owl. Err on the side of caution, give them space, and learn to read their body language.

Let them find their own food. Feeding owls is a cumulative problem. One person offering a live mouse once or twice won't make a difference (as long as they are not baiting the owl across a road), but there is no way you can know you are the only person offering the owl food, as owls can and do move, and social media can bring an influx of observers/photographers in a big hurry. A series of people baiting the same owl can cause owls to associate humans with food. Cumulative impacts are a very real thing due to birding websites and social media posts about owl locations. Assume every owl will encounter many observers/photographers, and let them find their own food.

Respect private property, fences, and signs. Farm fields are private property, whether labeled or not. Consider ALL property private unless it is posted as public. Owls have a tendency to make otherwise respectful people disrespect signs and private or restricted property to get the photo.

Avoid using artificial lights after dark. Although there are no published studies, simple experiments by Northern Saw-whet Owl banders show that owls banded at night will fly away faster and without issues if kept in the dark for five minutes before release, as opposed to owls exposed to lights before release. This indicates the night vision of owls is temporarily affected by lights. Our own

Short-eared Owl

experiments show that owls can see red lights and red lasers also, but not infrared lights.

Leave your dog at home. Owls have a natural fear of dogs. While some urban owls that are highly habituated to human activity may tolerate dogs, assume the vast majority of owls will be stressed by the presence of a dog. If you are observing human-habituated owls, observe the owl's behavior to see if dogs stress the owl.

Move slowly and keep your voice down. Fast movements and noises can stress owls. If you want to observe or photograph owls, move slowly and keep quiet to avoid scaring them.

Use your car as a blind when it's safe to do so. Owls are often more tolerant of humans in vehicles than outside of vehicles. Be sure to park on the CORRECT side of the road to avoid stressing other drivers (photograph/observe from the passenger seat), and turn your engine off. Hunting owls will thank you for the quiet so they can hear their prey better!

Leave your drone at home. Owls and other raptors often perceive a drone as a threat in their airspace, especially around a nest. This can be highly stressful and may provoke owls to attack the drone. Owls and other raptors can be injured by the blades of some drones when they attack them (and the owls may also damage the drone).

Think before you use playback. Owls often call back or come into (fly in to investigate) playback of calls of their species. This method may be used by researchers, but doing this purely for recreation needs to be carefully considered. It is illegal in national parks in the United States, because of the extreme disturbance, likely due to the high number of visitors. We strongly encourage people to avoid using playback in any public area to avoid repeated disturbance, significantly limit playback on private property, and never use it for rare species like Spotted Owls.

Think carefully before you share the location of an owl. Nesting owls are easy to photograph because the birds are tied to their location once eggs are laid. Some northern owls also become habitual about their wintering grounds and are easy to see. It is all too easy for these birds to be mobbed by photographers and other observers if locations are shared, especially on social media and birding websites. Although there is absolutely value in showing an owl to people who have never had the pleasure of seeing a wild owl, think very carefully about if the owl you have found is at risk of being loved so much it is functionally harassed by the public. A good practice is often to report the owl's location only at the county level or after the owl is no longer in the area.

Limit your time with the owl. Give wild owls plenty of privacy by keeping your visits short. Most species are trying to sleep for a big part of the day, and although they are normally awake for short periods throughout the day, they most likely aren't sleeping with you there. They are also trying to avoid being discovered by other bird species that will mob them, and your presence, especially if you flush one, increases their chances of being mobbed. Even habituated owls are going to be more relaxed when humans aren't around. Take into account how much time others are spending with the owl also, as the impact is cumulative.

Leave branches on trees. Owls have chosen their roost and nest locations in part because of the shelter from the weather and camouflage the location offers. Removing branches so you can get an unobstructed view of the owl is counter to the owl's wishes to remain hidden. Owls need to conceal themselves to avoid being mobbed by smaller birds, being eaten by larger hawks and owls, and feeling threatened by dogs, humans, and other perceived threats. Keep in mind that if you don't own the tree, it may violate laws or rules for you to remove branches.

Be respectful of other people. Share your binoculars, spotting scope, or camera view with others who may happen by without viewing optics, and share information with them. You may observe others who are not behaving ethically around owls. They may or may not be aware of how their behavior is impacting owls. Respectfully point out how their behavior is affecting the owl and model good behavior yourself. If they choose not to heed your advice, you may choose to tolerate their behavior, leave the situation, or report it to the authorities if it is illegal. Verbally assaulting people who are not behaving well or photographing them to smear them on social media will not improve the situation or help owls. Treat others as you would like to be treated.

apply to private land, with the added condition that you have the landowner's permission.

Placing, monitoring (on-site), and retrieving a trail camera can profoundly disrupt wildlife. That's why placing cameras at owl nesting and roosting sites should only be done by professional biologists for research purposes. For the rest of us, a camera discreetly placed at a water hole or water feature can reveal wonderful details about the nocturnal world, and some of the best such images I've seen come from people who maintain a trail camera at their backyard birdbath or similar water feature. It's a fun and educational way to learn more about the wildlife that shares your space. Moreover, as we create human habitat, we need to find ways to mitigate the loss of wildlife habitat, and providing water is one way to help do that. When I started maintaining a small water trough in the narrow tract of hardwood forest adjacent to my house, I installed an unobtrusive trail cam nearby so I could monitor wildlife activity from afar. The variety of creatures happy for a drink during the long summer surprised me, and included a Great Horned Owl, along with Cooper's and Red-tailed Hawks, many other birds, bobcats, coyotes, raccoons, opossums, native squirrels, and—much to my surprise—even a black bear that wandered through late one June night.

Drones

Trail cams used ethically are unobtrusive, but drones, or unmanned aerial vehicles (UAVs), are not. Today, millions of recreational (consumer) drones are registered with the Federal Aviation Administration. Federal, state, and local laws governing the use of recreational drones can hardly keep up with the burgeoning popularity of these devices. Drones are also increasingly used by scientists and wildlife management agencies to study many different animals. At the same time, scientists continue to learn how drones themselves impact the wildlife they purport to study and how recreational drones can disrupt wildlife.

Drone photography is immensely popular, but these small, quiet flying machines can alter wildlife behavior and place undue stress on birds and other animals. Harassment of wildlife is almost always illegal and is generally and broadly defined as any act that causes wild animals to alter their behavior in response to a human activity. But the explosion in drone use has forced lawmakers around the world to enact new laws to protect animals. Researchers may have legitimate reasons to use drone technology to study owls and other birds, but photographers who use drones should avoid owls at all costs. Even an owl that doesn't show outward signs of being disturbed may still suffer from undue, potentially harmful stress. Indeed, raptor researchers measuring stress hormones in birds have discovered that rehabilitated owls whose injuries prevent them from being released into the wild often suffer significant stress in the presence of humans; because of this epiphany, facilities that once used these captive owls for educational purposes are instead retiring them to quarters where human contact can be minimalized.

The author's trail cam experiment at a remote desert water hole revealed that a Great Horned Owl family resided somewhere nearby.

Recorded Calls

Another modern technology that has become all too prevalent among birders and photographers are recorded calls and songs of birds. Many species of birds will readily respond to vocalizations of their own species and sometimes other species, particularly during the breeding and nesting season. Recorded songs and calls are readily available on various apps and websites, and unfortunately, this technology has been widely overused and abused.

Playbacks, as these recorded calls and songs are known, are especially effective in eliciting a response from territorial birds during breeding season, and their use is so ubiquitous among bird photographers that a trained eye can often tell whether a close-up photo was obtained by use of playback to bring the bird out into the open and close to the camera. Photos of a diminutive songbird called the Ruby-crowned Kinglet are great examples: the small tuft of bright red feathers that give this bird its name are rarely seen, almost always concealed by the surrounding feathers. But play a Ruby-crowned Kinglet's scolding calls or songs during spring, and any nearby individual is apt to come out of hiding, excited and agitated—and prominently displaying the red crown feathers.

Researchers sometimes need playback technology, especially for studying nocturnal and crepuscular birds such as owls. For example, the tiny Flammulated Owl that inhabits coniferous forests of the West is listed as a sensitive species in the United States and a species of special concern in Canada. Flams, as they are often called, are

Identifying an owl, such as the diminutive Northern Saw-whet Owl, by its distinctive calls is entirely satisfying; we needn't actually see the owl for a successful owling mission.

vocal during the breeding season, but they are difficult to find because they are entirely nocturnal and typically breed and nest in montane (mountainous) forests. So, to monitor this species to detect any decreases in populations or changes in range, scientists rely on playback, but only to elicit an answering call from a Flam. In fact, Partners in Flight, a research and advocacy organization, instructs Flammulated Owl monitoring crews to use playback as follows: "You will spend a total of 10 minutes listening and calling for owls at each survey point. The 10-minute protocol will be split into five 2-minute intervals: two minutes of silent listening, and for each of the remaining 2-minute intervals, you will spend the first 30 seconds broadcasting, followed by 90 seconds listening. For the 30 seconds of broadcasting, play approximately 7.5 seconds with the caller pointed in each of 4 cardinal directions."

By minimizing the duration of playbacks, these owl monitors are locating owls by sound, but—critically—not trying to draw the owls toward the caller. That's where owl researchers and recreational owlers sometimes diverge. People using playback for nonscientific purposes sometimes try to draw an owl into close proximity so they can actually see the bird, which puts a small owl at risk of becoming prey for a larger species. Other than Great Horned Owls and probably Snowy Owls, all North American owls are subject to predation by other owls and other large raptors. Even big owls such as Great Gray Owls and Barred Owls are sometimes killed by Great Horned Owls, and small species are in even more peril.

Ethical owlers understand that *hearing* a species, such as a Flammulated Owl, is good enough, whether accomplished via a very brief playback, or better still, through enjoying a quiet evening in the forest, ears attuned to the possibilities. At the least, recorded calls and songs should be used judicially with any species of birds, but in my opinion, not at all with owls. Owls have no choice but to vocalize to attract and communicate with mates, offspring, and others of their species. Fooling an owl into vocalizing by playing its calls places the bird in unnecessary danger of being located by a bigger owl, or of having its nest raided while it's busy investigating the source of the calls. Owlers relying on playbacks to locate the birds they claim to admire risk violating the overarching ethic to do no harm. As the International Owl Center says, using playback "purely for recreation needs to be carefully considered," and owl experts and authors Pat and Clay Sutton are even more succinct, saying, "Never use tapes or imitations during an owl's breeding season unless you are part of a legitimate organized survey."

Thermal Imagery

Thermal imagery is the latest technology to arrive in the bird-watcher's tool kit now that thermal monoculars and binoculars are available in much better quality and at lower prices than just a few years ago. These optic devices literally allow you to peer into the night, revealing animals as warm colors against dark backgrounds. Under optimal conditions, with a thermal imaging monocular, you can find an owl sitting in a tree very easily, and from a distance. This technology is discussed in more detail later, and while extremely useful to owlers, thermal imagers can lead to unethical bird-watching behavior: by detecting the bird's body heat (infrared wavelengths), a thermal imager allows you to find an otherwise hidden owl. That's a boon. But when birders then use this intel to try to get close to the hidden owl, they may cross the line into unethical behavior. Remember, owls hide to avoid predators both to themselves and their nests/young, and to aid them in hunting unsuspecting prey. We never want our enthusiasm for seeing owls to interfere with their natural behaviors.

This soggy Great Horned Owl must have been hunting the abundant shorebirds or waterfowl at an expansive shallow-water marsh nearby. The author noticed the bird tucked into a grove of willows, trying to dry in the morning sun.

How to Find Owls

I've seen lots of owls because I've spent lots of time outdoors. As prosaic as that may sound, it's true. The easiest way to see owls is to immerse yourself as often as you can in places where owls live and keep your eyes and ears open—regardless of whether your primary objective is to see owls.

However, I have friends who spend lots of time outdoors and haven't seen lots of owls. They are focused on other activities or other things. They miss a lot of owls (and other creatures and curiosities).

Finding owls by accident isn't always accidental. It's just a matter of awareness. Owls of one kind or another occupy just about every imaginable habitat type, so virtually any activity that takes you outdoors gives you a chance to see an owl, but they are seldom obvious. Perhaps more than any other form of bird-watching, owling teaches us to be better observers, to be more aware of our surroundings.

I call it accidental owling, but secondhand owling might be more apt: seeing (or hearing) owls while engaged in other pursuits isn't necessarily accidental. An owl might be too obvious to miss, but an owl might be easy to miss while you focus on something else, such as hiking, walking the dog, fishing, hunting, rockhounding, foraging for mushrooms, beachcombing, photographing, picking wild berries, even weeding the yard at dusk. I've seen owls during all those pursuits. And that's just me. Whatever hobby or task takes you outdoors provides you the opportunity to see owls, by accident (because the owl was obvious) or by secondhand owling (because you've learned to multitask, to be observant while outdoors).

Keen observation is also critical to purposeful owling, in which hearing and/or seeing owls is the mission. Owls are secretive by nature and physically cryptic, with their excellent camouflage, but they share these traits with many kinds of birds. Unlike most birds, however, all but a few species of owls are nocturnal or crepuscular, active after dusk and before dawn. By day, they hide, and therein lies the challenge—and the reward when you actually do see an owl. Owling requires extraordinary observation skills, but also attention to detail: the more you know about owls, the better your chances of finding them. For owlers in North America, the identification guide in this book is your first step to becoming an owler, part of the step-by-step owling strategy outlined in this chapter.

Owling is both a daytime and nighttime activity, depending on whether your goal is to see owls or hear owls, or perhaps a little of both—and results vary by species. Naturally, large owls are easier to locate and see than small owls; owls that hunt during the day are easier to see than nocturnal species. In the far north, provided you visit appropriate locations,

Bird-watchers are generally quick to take note when an owl is found in the same place day after day, such as this Snowy Owl, and there's nothing wrong with getting in on the thrill of visiting the site, if we always keep the owl's best interest in mind—keep your distance, be quiet and unobtrusive, and remind others to behave likewise.

Snowy Owls and even Northern Hawk Owls are easy to see because they are active in daylight (they have little choice during the summer when daylight persists for all or most of 24 hours). Farther south, Short-eared Owls often hunt around dawn and dusk, and they search for prey on the wing, coursing over marshes and prairies. They are easier to see than species that hunt after dark. On the other hand, many small species of owls hide in holes, crevices, and other such crannies during the day, often completely out of sight.

Seeing owls is fun—they are cool birds! When owls take up temporary or permanent residence in obvious places, they tend to garner a lot of attention from onlookers. This is especially true of less-common species. A Great Horned Owl that routinely hangs out in an obvious, accessible location won't attract nearly as many birding enthusiasts as a Long-eared Owl that makes an appearance at such a place. In fact, some years ago near where I live, a Long-eared Owl chose an expansive grove of trees as its favorite day roost—and this location just happened to be along a boardwalk trail through part of a national wildlife refuge. As soon as a few birders discovered the owl, word got out within the regional bird-watching community and the owl quickly began attracting increasing numbers of onlookers. Most people were well behaved, and fortunately for the owl, that boardwalk trail passes through a wetland—there was little danger that overeager camera-wielding birders would try to approach ever closer.

That's not always the case when an owl is discovered. A few years later, a similar crowd began assembling around a Short-eared Owl at a refuge in Washington State. Word had spread via social media and other means, and soon the roadway through the refuge was swarming with vehicles. The refuge rule stating that drivers must limit stops along the roadway (for viewing and photographing wildlife) to five minutes went right out the window, so to speak, resulting in congestion that could only be described as a traffic jam. A fine line can separate wildlife viewing from wildlife harassment, and some people crossed that line.

But if we always keep the best interest of the owls in mind, visiting a site where an owl has been reported is a great way to see a species new to you, to capture great photographs, to observe owl behavior, and to meet like-minded people. Every winter, owls of the far north—especially Snowy Owls, but also Northern Hawk Owls and Boreal Owls—fan out southward, and some end up in unlikely places. When a Snowy Owl spends part of its winter in the Lower 48, it almost always attracts great fanfare because the occurrence provides an opportunity for owl fans to observe a starkly beautiful owl they'd normally have to travel significant distances to see.

Years ago, when a Snowy Owl spent the winter in remote Harney County, Oregon, I was lucky: at the time, my friend Tim Blount lived in Burns, Oregon, and told me about the owl. I waited until the excitement subsided; Tim reported to me that the owl, usually seen perching on the ground or on wooden fenceposts on private ranchlands along a paved back road, had drawn lots of admirers, most of them driving for many hours from places like

Portland and Eugene. But two weeks later, the fervor had waned, and the owl was still there. I was able to pull off the road at a respectful distance and capture a few images with a telephoto lens on a sunny winter morning with nobody else around.

Viewing unusual species of owls is always thrilling, but again, it's a lot easier with large owls and owls that are active during daylight. Many owl species are just flat-out hard to find with your eyes, and that's where nocturnal owling gets exciting. Most owls have a variety of vocalizations, and many of their calls are haunting and exotic. Sitting quietly outdoors and listening to owls—whether the familiar deep hoots of the Great Horned Owl, the spine-tingling screams of the Barn Owl, or the high-pitched echoing toots of a Northern Saw-whet Owl—is exhilarating. And you can hear owls almost anywhere, from expansive wilderness to urban parks, campuses, and even cemeteries (appropriately, given the owl's historical associations with death, darkness, and the occult). If you find a mix of open spaces and woodlands, you stand a good chance of finding owls (other than open-space-shunning species like the Spotted Owl).

Where I live, not far beyond city limits, in a neighborhood with small fields and small woodlots, I never tire of the nightly serenade of Great Horned Owls between about December and March. During winter, they perform duets, the male with his deep hoots and the female with her higher-pitched replies. Some nights three or four Great Horned Owls join the choir. And then, occasionally, comes the piercing scream of a Barn Owl, and sometimes, from the oak-clad hillside nearby, Barred Owls announce themselves with their iconic and exotic *who-cooks-for-you* calls. A few times over the years, I've been lucky enough to be outside—gathering an armload of firewood, or taking the dog out, or bringing in the hummingbird feeders in advance of subfreezing weather—when the enchanting bouncing-ball song of the Western Screech-Owl echoes through the cottonwoods and maples.

Obviously, I reside near good owl habitat, but such places abound. If you live in the city, rest assured you needn't travel far to find the places owls live. And the more you learn about these birds that haunt the night, the more of them you'll see and hear, whether owling or other activities take you outside. My eight-step plan as outlined here is your key to successful owling, and all eight parts are easy and fun to implement.

Step 1: Know Your Owls

Learn which owl species live in your area or places you visit and learn to identify them by sight and by sound. If you are new to birding in general, and to owling, don't worry—you don't need to memorize every detail about

The author happened upon this Long-eared Owl in a remote expanse of desert scrub—not surprising for a bird that frequents open spaces.

each of the 19 species of owls found north of Mexico. Start with the species that live where you want to look for owls. Check the range maps in chapter 2 for each species: for example, if you live in New England and want to see some of the species in your area, you can rule out more than half of those 19 species. Moreover, of the species that do live in New England, consult the "status" subhead for these owls to determine which are most common and which are rarely seen in the region.

Another way to use ranges to help you find owls is to single out a species you want to see and then visit places where that species is most common. This is where an owl's status works well in conjunction with its range: for example, you stand a better chance of finding a particular species of owl in a general location where it is rated as common than a place where it is rare. And some of our 19 owls occupy very small ranges north of Mexico, the best examples being the Ferruginous Pygmy-Owl and the Whiskered Screech-Owl, both of which are found only in specific locales in southernmost Arizona.

Step 2: Understand Habitats and Habits

When you know the range of an owl species, you can then further enhance your chances of seeing or hearing that species by learning what kind of habitat it prefers and how it behaves within that habitat (its habits). Most (but not all) owls in North America are birds of the edges—they need dense cover for roosting and nesting but more open space for hunting. Meadows, fields, marshes, roadways, parks, steppe, and other such open environs that border woodlands, brushy or tree-studded riparian zones, or other good cover are prime habitat for many owls. Even open-understory woodlands, where owls have lots of places to perch, provide the open ground they need to hunt rodents and other prey.

Our most familiar species—Great Horned Owls, Eastern and Western Screech-Owls, Barred Owls, Long-eared Owls, and Short-eared Owls—prefer to hunt open and relatively open spaces, but they spend their off-hours hidden in nearby cover. Some other species are adapted to open habitats, while others live within the forest and shun open spaces. Burrowing Owls frequently inhabit treeless expanses of desert and steppe; Spotted Owls live within undisturbed mature forest. If you want to see a Northern Spotted Owl in the Pacific Northwest, you need go deep into the shaded conifer forests of the mountains; if you want to see a Burrowing Owl in the Pacific Northwest, you must visit the open high deserts and plains east of the Cascade Range.

Happily, for owlers, some owls are habitat generalists. Great Horned Owls are ubiquitous, able to thrive in a wide range of habitat types. You can find them in the mountains and in the lowlands, in the desert and in the woods, in the city and in the most remote places imaginable. Barred Owls and Eastern Screech-Owls are likewise highly adaptable and frequently found living largely in harmony with human neighbors in urban and suburban areas that provide them with a mix of open spaces and roosting cover. Barn Owls earned that name by their proclivity for roosting and nesting in human-made structures, particularly unused or little-used buildings with easy fly-in and fly-out spots such as open doors, fallen sections of walls, open ventilation ports, old window frames, and so forth. Other species of owls are also apt to roost and sometimes nest in old structures such as seldom-used or abandoned barns and sheds. No matter where your

Check abandoned and seldom-used buildings for owls and owl signs. Some species, such as Great Horned Owls and Barn Owls, routinely roost and even nest in human-made structures.

owling adventures take you, check out any such structures that are accessible to the public or by permission from the landowner. Sneak carefully up to an opening and scan the interior of the building, especially ceiling beams, crossbeams along the upper walls, and any other potential perches. If no owls are present, check the ground/floor for pellets, droppings, and feathers (see Step 4).

Other owl species are more habitat-specific. The Elf Owl of the deep Southwest, for example, favors riparian vegetation or saguaro cactus stands in lowland deserts. Likewise, the Boreal Owl is almost always found within conifer forests mixed with at least a few stands of birch, aspen, alder, or similar montane hardwoods.

Within appropriate habitat, understanding an owl species' habits can further tilt the odds in your favor. For example, Short-eared Owls often hunt in the wee hours of dawn and dusk (and sometimes in daylight). They hunt over open spaces, such as plains and marshlands, and often roost at the edges of those habitats. So, if you can find likely habitat and position yourself for a wide field of view at dusk, you stand a better chance than if you tried to find a Short-eared Owl roosted during the daytime. Tiny Flammulated Owls, on the other hand, are strictly nocturnal and spend the breeding season in western conifer forests at mid- to high elevations. You may never see one, but you stand a good chance of adding this species to your life-list by spending a springtime night or two within appropriate habitat and sitting quietly in the dark (huddled against the montane chill, as I know from experience) from dusk into the night.

Without question owls are much easier to find by sound than by sight. However, they tend to be most vocal during the winter and spring breeding season; thereafter, some species are almost entirely silent. Large owls are more obvious than small owls, and less prone to secreting themselves in crevices and holes during daytime. Purposeful owling to find and see smaller species is perhaps the ultimate bird-watching challenge, a needle-in-the-haystack quest, to be sure. In heavily forested regions, finding a forest species—a Northern Saw-whet Owl or

Flammulated Owl, for example—by sight is a most fortuitous stroke of luck. That's why such sightings are almost always accidental to other activities, including at times during bird-watching expeditions aimed at seeing other species. But in open country, you can narrow the search more easily by focusing on places where owls can find daytime hideaway roosts, such as isolated groves of trees and treed stream corridors.

Step 3: Learn Owl Calls

The little Flammulated Owl is a great example of why owlers need to learn owl calls. Flams are difficult to find with your eyes. They are nocturnal, tiny, cryptic, and uncommon. But their surprisingly deep and resonant hooting calls carry a considerable distance and are diagnostic.

Flams are one of the owl species that hoot, but people are often surprised to learn that many owl species don't hoot. In fact, most owls have a variety of calls. The Boreal Owl of northern forests utters a rapid continuing series of toot calls, which might fit the description of hooting, but they also use insect-like twittering calls, whistling-type calls, and even exotic-sounding wailing calls. And Boreal Owls are typical of owls in their variety of vocalizations. Some owl calls are downright eerie, such as the screaming shrieks of a Barn Owl and the spooky winnowing of an Eastern Screech-Owl. The deep resonant hoots of the Great Horned Owl appear in movie and television soundtracks, but the spine-tingling calls of other species would probably serve better in many a horror flick.

Learning to identify owls by their calls is challenging but a great deal of fun—and critical to successful owling. Describing owl calls, as I've done in chapter 2, is highly subjective, but it is a starting place. However, there is no substitute for hearing the calls. Recorded owl calls, from all the species in North America, abound on the Internet and on birding apps.

My favorite source of these recordings is a website called xeno-canto (www.xeno-canto.org), which catalogs recordings of birds from all over the world. This free-use site allows you to search for a species and then select from all uploaded recordings so you can listen to them on the spot. Multiple recordings of each species are especially useful because owl calls often vary geographically: for example, on xeno-canto, a Long-eared Owl recorded in eastern Washington State has a sharper hooting call than the more deliberate hooting of a Long-eared Owl recorded in Arizona. For each species, the xeno-canto website provides a list of recordings and a map that pinpoints the general location where the recording was captured. Click on an icon in the list or on the map to hear the vocalizations. Moreover, the list of recordings tells what kind of vocalization each recording preserves—songs, calls, alarm calls, male advertising calls, and many more. This resource is engrossing and highly instructive.

As you dive into the world of owl vocalizations, keep Step 1 (Know Your Owls) in mind: you don't need to memorize the calls of each of the 19 owls found north of Mexico; you need to learn the calls of the species found where you go owling. Moreover, if you are scouting locations for potential owl sightings, visit at dusk (especially during winter and spring), sit quietly, and listen for 15 to 30 minutes. If you hear one or more owls, you can reasonably expect that the birds are day-roosting in the area.

Step 4: Do Your Homework

Steps 1 through 3 help you learn more about owls, particularly those in the places you go owling. Step 4 helps coagulate all those details into an actionable plan to find owls, by sight or sound. Many online and hard-copy resources will help you dial in locations and times to see owls, making your owling adventures harbor at least somewhat better odds than a shot in the dark. For example, bird checklists covering specific sites, such as wildlife refuges, state parks, national parks, and wildlife sanctuaries are frequently available on-site and often online. They are ubiquitous and helpful.

The most useful checklists include information about seasons and relative abundance, telling you what time of year a certain species occupies the site and how common that species is at the site. For example, a checklist for remote Great Basin National Park in Nevada tells you that nine species of owl inhabit the park, and that seven of them are year-round residents, two are breeding-season residents, only one is listed as common, another as uncommon, and the rest as rare. Or perhaps you are headed to Itasca State Park in Minnesota—an Internet search for "Itasca State Park bird checklist" quickly reveals that the Minnesota Department of Natural Resources provides an online checklist for virtually all state parks, and you might decide to visit Itasca for a chance at seeing or hearing any of six different species of owls during the spring.

eBird, a website and app from Cornell University Lab of Ornithology, is another valuable research tool for owlers. On the app or at www.eBird.org, click on "Explore," which takes you to several options: you can explore regions, birding hotspots, species, and more. The "Species Maps" section is especially useful because it catalogs sightings reported by observers and marks them on a map, sometimes at specific locations and sometimes at broader general areas (more on that in a moment). So, if you want to search for places where you might stand a chance of seeing or hearing a certain owl species, you type the name into the "Enter species name" box, then click on the name in the drop-down menu that pops up. This brings up a zoomable map colored with pink and purple rectangles; the more you zoom in, the more specific the location and shade of the rectangles—deep purple for more reported sightings at that location, light pink for few reported sightings. Those purple rectangles are your best bet for choosing locations that might provide hope of seeing or hearing your target species.

For most birds, you can continue to zoom in to the species map until the purple and pink rectangles are replaced by blue and red marker pins that show specific locations. But this feature is not available for owls. That's because Cornell keeps the best interest of owls in mind by including them on its "Sensitive Species" list: specific-location owl sighting reports by observers can lead to people mobbing sites where an owl has been found, and such behavior by people is not healthy for owls. So, for owls, eBird does not allow reporters to list locations by name, but those purple boxes are still great clues to general locations where you might find the owl you seek.

There is a somewhat useful work-around on eBird: you can check species reported by birders for specific places by using the "Explore Hotspots" feature within the "Hotspots" tab. These lists give you a snapshot of what other visitors have seen; for helping to find opportunities to see or hear owls, this feature works best if you can narrow the location as much as possible. For example, the hotspot feature lists about three dozen locations within Arizona's Saguaro National Park (SNP). If you're planning a multiday visit to this national park, taking time to study the lists from all those sites might be worthwhile. Begin by selecting one site within the list of locations for SNP, let's say "Widlhorse Canyon area." When you click on that location, you'll redirect to a map and a little boxed inset that shows the number of species and the number of checklists for that area. In that box, you can click on "View Details" to see the species, but better yet, click on "Bar Charts" to bring up a list of sightings organized by time of year (and at the top you can select a date range to eliminate older sightings). Within the bar chart for Wildhorse Canyon area, three owl species are listed: to the right of each species listed, green squares show the months when that species has been reported. Immediately to the right of a listed species, click on the first of the two blue icons to bring up a map with pointers showing specific places and sighting details. This process can be time-consuming but also highly informative—another tool to help you narrow the possibilities for owling locations.

In addition to using eBird, join Facebook groups dedicated to bird-watching in your state or region. Owls that attract a lot of viewers may be reported with location by members of these groups, but most people have learned not to hotspot their personal owl sightings (and most Facebook birding groups discourage or even forbid hotspotting). Nonetheless, sometimes a personal message to someone who has reported a sighting can lead to an invitation or a location report, not to mention perhaps a new bird-watching friend. In addition to state and regional birding Facebook groups, check out the numerous Facebook pages and groups dedicated to owls in general (including the educational and entertaining Facebook page of the Owl Research Institute).

Finally, many state and regional birding organizations maintain rare bird alert programs and frequently updated lists of bird sightings reported by list subscribers. Often these lists are available via email, and many are accessible via the search function at www.aba.org/birding-news, a page on the American Birding Association website.

Step 5: Learn the Sign Language

No, owlers don't use some secret handshake or furtive sign language (although they do sometimes meet in dark, recondite places). Rather, owls themselves leave signs of their whereabouts, and learning to identify and decipher signs of an owl's presence, or potential presence, helps immeasurably in consistently locating these birds.

Look down to see up: remember those regurgitated pellets we learned about in chapter 1? Keep a sharp eye out for them, and even more so for whitewash (sometimes called guano); the two are often found at the same site, places where an owl has been roosting, eating, or nesting. Lots of birds excrete white, runny poop, but it builds up over time beneath perches that are used repeatedly. And owls are among the many species that reutilize favorite roost sites. Whitewash is perhaps the single most important sign to look for because it tends to be the most visible, often painting streaks and splotches on tree trunks and limbs where an owl spends time.

Learn to look for whitewash and to differentiate it from whitish sap. As you stroll slowly through woodlands (where many species of owl roost during the day), examine the trees and limbs. If you see whitewash, stop and study that tree with binoculars. If you don't see an owl after a thorough search, move in slowly to study the ground to look for pellets and additional whitewash. If you find whitewash on the ground and/or on the tree above, you may have found an owl hangout; if you find regurgitated pellets, you've probably found an owl hangout. Also look for feathers; birds routinely lose and replace feathers, and owl feathers often begin to accumulate beneath favorite roosting sites and nest sites. Again, though, always examine the trees or other roost sites (cliffs, bluffs, dense shrubs, buildings, etc.) from a distance with binoculars before moving closer to search the ground; if you do see an owl, keep your distance and enjoy the sight through binoculars.

Also look for holes, crevices, and cavities. Most small owls roost (and nest) in crevices and cavities, including woodpecker nest holes and sections of trees excavated by woodpeckers searching for food, scar holes left by broken branches and trunks, and forks where limbs meet trunks. Roosting habits vary by species, so this step-by-step guide to owling consists of steps that work together: learn the roosting habits of the species you might find. For example, you won't find a Great Horned Owl in a woodpecker hole, but they will roost in a wide range of places, from nooks and crannies in cliffs to large trees to abandoned buildings; come springtime, train your binoculars on nests made by Red-tailed Hawks and other raptors, ravens, magpies, herons and egrets, and even squirrels—Great Horned Owls frequently commandeer stick nests built by other species.

And speaking of other species, owls are prone to being mobbed if they are discovered by other birds, and squirrels may also help sound the alarm with excited chatter. So, listen acutely for agitated birds when you are owling during daytime, particularly in woodlands. Common and vocal owl mobbers include jays, crows, chickadees, titmice, kinglets, nuthatches, wrens, juncos, and others. If you hear such commotion—particularly several species of songbirds chattering and scolding in excited tones called scolding calls and alarm calls—try to zero in on the location, but keep your distance from the scene so as not to further disturb a poor owl that is already suffering the indignity of having been found by songbirds.

The author noticed the whitewash on these rocks and returned to the spot several times before finally seeing the occupant of the perch one morning—a Long-eared Owl.

Finally, some other bird species are potentially predictive of the presence of certain owls because they occupy the same habitat. For example, the Northern Harrier, a kind of hawk, hunts for rodents in the same open habitats favored by the Short-eared Owl. Harriers hunt by day, and Short-eared Owls typically hunt the same marshlands, fields, and prairies around dawn and dusk. Red-tailed Hawks often occur in the same habitat as Great Horned Owls because these hawks nest in tall trees and frequently hunt via the perch-and-pounce strategy, surveying the land beneath favorite perches and watching for rodents below. Great Horned Owls hunt the same way but at night, between dusk and dawn. If you live in the rural West and your birdfeeders routinely attract bird-eating Sharp-shinned Hawks or Cooper's Hawks, keep your eyes peeled for Northern Pygmy-Owls, which also hunt smaller birds by day and sometimes prowl backyard birdfeeder stations.

Step 6: Tilt the Odds

One of the best ways to become an owler is to spend time with other owlers, particularly experts who lead owling expeditions. Check with chapters of the National Audubon Society, birding clubs, wildlife sanctuaries and preserves, nature centers, state parks, and state wildlife agencies in your area to find out if they are planning any owl outings. The International Owl Center in Minnesota, for example, offers a full slate of owl prowls each fall and winter, and such events are routinely offered by other groups. "Owl prowl" has become a favorite title for guided owl expeditions, which are widely available but also popular enough that making reservations well in advance is wise. Many of these events occur—understandably—at dawn, dusk, or after dark, and many include educational components. They usually take place in winter so that owls are not unduly disturbed during the spring nesting and fledging season.

But doing your daytime owling in late fall and winter also tilts the odds in your favor because deciduous trees have dropped their leaves, making perched owls easier to see. An owl hiding amid foliage is extremely well camouflaged, so when the trees are bare of leaves, you stand a much better chance—look closely at branches near tree trunks and any crevices or cavities on those trunks and large branches. Big owls like to huddle close to trunks, and smaller owls often hide in crevices.

Owl prowls and similar owl-specific outings are great ways to learn a lot about owls, but more generalized bird-watching trips and tours are also valuable, especially for seeing a particular target species. Again, check with Audubon Society chapters, birding clubs, refuges, and nature centers. For example, in my home state of Oregon, a tour of the Malheur National Wildlife Refuge (perhaps with the Portland Audubon Society) would be a very good bet for seeing Burrowing Owls. Species-specific owling can be highly successful if you go to the right place at the right time, and options abound for owlers to pursue the single-species strategy, from guided professional tours to do-it-yourself trips after you've done the necessary homework.

Also, consider helping the people who study owls. Some owl research projects involve monitoring populations, which frequently requires scientists to head out into the night to listen for owls. They often need volunteers; Birds Canada, for example, recruits more than a thousand volunteers each winter to help with its comprehensive late-winter/early-spring nationwide nocturnal owl surveys. Other surveys (all around the world) might target just a single species within a small region. Regardless of the scope of their projects, owl surveyors often need help. Google "volunteer for an owl survey" to get an idea of where and when you can get involved. These surveys are great opportunities to learn about owls, hear and possibly see various species, and contribute to owl research as a citizen scientist.

A guided owl prowl hike begins before dusk at Virginia's False Cape State Park. Joining an owl prowl or similar event is a great way to meet other owl fans and learn from owl experts.

Step 7: Assemble Your Owling Tool Kit

A keen sense of observation is the most important asset for owlers, with patience being a close second. Seeing owls often requires a significant investment of time and effort—and you must embrace failure because owls don't want to be found. But even if you don't find an owl, those hours you spend quietly searching woodlands and groves of trees and other "owly" places are never wasted because this treasure hunt almost always yields many other pleasant and intriguing sightings. Between dawn and dusk, you are sure to see many other kinds of birds and other wildlife, making a good pair of binoculars the one critical tool in your owling kit. For nocturnal owling you will rely primarily on sound, listening quietly for owl vocalizations. But carry your binoculars nonetheless to zoom in on any owl sighted during crepuscular hours.

Also consider thermal image devices, such as a thermal monocular, which allows you

Hidden in a broken-branch scar, an owl is well camouflaged (left), but a thermal imager easily reveals the bird's heat signature (right), and a zoom-lens photo reveals the cryptic bird as an Eastern Screech-Owl (bottom).

to literally see in the dark, at least in terms of finding owls and other birds secreted away in the concealing foliage or hidden by the cover of darkness. Even in daylight (especially in cold weather), thermal detection optics can find the heat signature of living creatures, making them glow in the dark through the device's viewer. Thermal imaging is a form of night vision but a different technology than optics that allow you to see better at night by amplifying ambient light. Thermal imagers interpret infrared wavelengths into a visible spectrum—all objects that have a temperature radiate infrared wavelengths, and warm-blooded creatures such as birds emit lots of infrared. Our eyes cannot see into the infrared spectrum, but a thermal imaging device can do that for us, displaying a bird's infrared wavelengths as bright colors. Thermal imagers work especially well in cool or cold environments.

Thermal imaging monoculars are ubiquitous, so much so that trying to decide which model to buy can be overwhelming. Weeding through all the specs for different models, look for the two categories that matter most: sensor resolution and field of view. As with cameras, the higher the sensor resolution, the cleaner the image, but thermal imagers don't need the high pixel density of cameras because they pick out heat signatures—blobs of bright colors—rather than fine details. Like standard binoculars, thermal imagers come in a range of field-of-view options, but generally narrower is better. With a narrow field of view, you see better detail, but you must scan more to see more of the environment than with a device that has a wide field of view. You may suffer sticker shock when you shop for a thermal image monocular, but even as the technology in these devices continues to improve, the prices seem to be moderating.

Beyond binoculars and perhaps a thermal monocular, your owling tool kit should include a flashlight or headlamp for nighttime forays.

Use it judiciously to prevent stumbling and other hazards, and to study maps. Never use artificial light to illuminate owls—they are trying to hide after all, both to gain the edge on potential prey and to avoid becoming prey for something bigger, usually a bigger owl. Ideally, your headlamp or flashlight should have a red-light setting—the red light is less disturbing to animals and provides all the light you need for reading a map, checking a field guide, or tying a boot lace.

Step 8: Get Out There and Have Fun!

You've studied your owls, done your homework, and learned more than enough to anticipate successful owling. If you're heading out on a guided owl prowl, the leaders will choose the locations, but it's up to you to be prepared. Most owl outings occur during late fall and winter to avoid interfering with owls during their late-winter/spring nesting and chick-rearing season. So, dress for the cold, especially if you are owling at night. Remember to pack your tool kit, especially binoculars, and perhaps a notebook and pen.

On the other hand, if you're planning your own owling expedition, maybe with the family or with friends, remember the old real estate axiom: location, location, location. Owls are everywhere. Almost. But they can be hard to find, even in prime habitat. Your research will aid in your choice of venues, perhaps leading you to well-known owl hangouts. But—again, with habitat in mind—you can also make an educated guess. Choose a few tracts of owly-looking habitat and scout them as thoroughly as you can in advance of an owling expedition. In a suburban or urban setting, maybe you know some large parks, cemeteries, or college campuses with plenty of trees (especially groves of trees) and ample open spaces; maybe a riverside trail winding through tracts of riparian woodlands. In more rural settings, scout for owl habitat (keeping your local owl species in mind) and make notes on a map. Apps that show property ownership (e.g., OnX, BaseMap, LandGlide) are helpful in not only discerning public from private lands but also in determining who owns parcels in case you want to seek permission to make owling forays on private property. From remote regions to suburban settings, owling possibilities abound.

Most owls are typically easiest to hear and see in the crepuscular hours, particularly when they begin hunting around dusk, but sometimes also when they return to their day-roosts before dawn. Many species are vocal well into the night, especially during winter, so the easiest way to locate owls is with your ears. But successful nighttime owling begins during daylight: plan an owling route that includes several or even numerous stops. Drive the

Do you see the owl? Owls are cryptic, and small owls are especially adept at hiding in plain sight. This Western Screech-Owl blends remarkably well with the tree bark.

route during the day to locate safe parking areas and jot them down on a paper map or notepad. Then, come nightfall, stop at each location, be very quiet, and listen for owls for perhaps 10 minutes before moving on to the next stop.

Hearing owls is one thing. Seeing them is quite another. Although the crepuscular hours are ideal, because owls are often vocal then, as well as active, those lowlight bookends of the day are narrow, fleeting windows of time. Seeing owls is akin to treasure hunting—the odds are stacked against you, but it's ever so rewarding when you succeed. Large owls are easier to see, of course, and not only because of their size. Big owls often sit at the base of a large branch, in a fork in the trunk, or even in broken snags that offer concealing jagged edges. Small owls sometimes roost in similar places but also happily take advantage of holes and crevices. Screech-owls are renowned for their ability to blend in with the tree bark while huddling into a scar left by a broken-off limb or any similar crevice. But other small owls roost in similar places. Remember, late fall and winter proffer the benefit of trees bereft of their leaves, making trunks and limbs easier to see, at least on deciduous trees.

Unadulterated luck may reveal a hidden owl: you looked in just the right place. But systematic searching yields better or at least more consistent results. Above all else, go slow! When you walk quietly through a grove of trees, study each trunk carefully, from the ground up; look closely at large limbs, especially where they branch off from the trunk. Then when you walk—slowly—past a stand of trees, examine the other side of the trunk; scrutinize any crevices, such as indented scars left where limbs have broken off. Always look for whitewash, which is one of the key give-aways to an owl's roost. Keep a sharp eye out for stick nests made by hawks and other birds (some owls commandeer them for their own nesting needs); when you spy holes in trees, quietly examine the ground beneath (binoculars can be handy for this). Owling by day is certainly challenging, but finding a hidden owl on your own is remarkably satisfying and you'll encounter all kinds of other creatures along the way, making those days of fruitless owling rewarding nonetheless.

4

Give a Hoot

Help the Owls

Despite popular legend, Samuel Colt's famous revolver didn't tame the American West. Barbed wire did.

The brainchild of Illinois farmer Joseph Glidden, barbed wire—aptly called the devil's rope—took the country by storm in the 1870s. By then, westward migration to many areas had slowed because aspiring farmers had no adequate means of protecting their crops from livestock and wild ungulates. Any passing herd of cattle or bison could prove ruinous to crops planted through backbreaking labor on the frontier. But barbed wire changed everything, and westward migration exploded. As historian Roy D. Holt explains, "More white settlers moved further westward in the eight years after affordable material for fencing was introduced than in the 50 years prior to that."

In 1880, just six years after Glidden invented the fencing that would earn him a fortune, his factory in DeKalb cranked out 263,000 miles of barbed wire, enough to circle the Earth 10 times—and a fraction

Barn Owl

of today's annual worldwide output. Some of that output injures and kills wild creatures every year.

Barbed wire is so dangerous to wildlife that decades ago, the federal government initiated efforts to retrofit barbed wire fences in regions populated by pronghorns (aka pronghorn "antelope") in the American West. Pronghorns are built for speed, not jumping ability, so they crawl under fences rather than leap over them like deer. Barbed wire severely hinders the ability of pronghorns, which teetered on the brink of extinction early in the 20th century, to move safely across the range, particularly during seasonal migrations. At the same time, ranchers and range managers deem barbed wire necessary to contain livestock and prevent grazing animals from ranging beyond allotted grazing areas. So, the retrofitting program seeks to remove the lowermost (nearest the ground) barbed wire (most rangeland fences have three strands of wire) and replace it with smooth wire, which minimalizes the risk to pronghorns when they need to crawl under a fence.

Pronghorns are now much safer in the West, yet barbed wire fences still kill and injure uncountable numbers of birds and mammals every year, and owls are especially vulnerable, a point driven home to me one autumn morning when a buddy and I were driving a winding backcountry road through a scenic canyon in north-central Oregon. As we rounded a bend beneath towering basalt ramparts, I suddenly asked my friend to stop and back up; I had seen what appeared to be a bird hanging from the dilapidated barbed wire fence paralleling the roadway.

It was a Western Screech-Owl that had obviously flown into the fence during its nighttime hunting forays, and was impaled at the shoulder under one wing, blood staining its feathers. Despite the wing injury, the owl seemed alert but unable to extricate itself from the iniquitous barbed wire.

Our fishing trip became a rescue mission. Using gloves and great care, we freed the owl from the fence, then placed it in a big paper sack on the back seat of the truck. We backtracked, driving out of the canyon to return to the main highway, then to the nearest small town to find cell reception so I could search the Web for animal rescue facilities. We were surprised and elated to discover that a raptor urgent care/rehabilitation center was a mere two hours distant. I called ahead, and the facility manager urged us to bring the little owl in. Weeks later, upon the owl's full recovery, we were tasked with returning it to its home and we gleefully watched the bird fly up into a dense grove of trees and shrubs in a narrow side-canyon, classic Western Screech-Owl habitat.

The incident was my first experience dealing with raptor rehabbers and many years later was impetus to interview professional wildlife veterinarians and rehabilitators to gain an understanding of the human-induced perils that owls face and learn how we can mitigate the dangers. Sure enough, my interviews revealed that barbed wire is a major threat to owls. Hunting in the dark or in dim light, they often cannot see the strands of wire, and because they primarily snatch prey from the ground, owls are in danger of entanglement.

While deforestation and loss of habitat all over the world are the biggest threat to owls in general, barbed wire, and other causes of entanglement, are owl perils we can easily do something about. And the same is true of many other dangers that lurk for owls—collisions with vehicles are especially onerous, pesticides (especially those used to kill rodents) indiscriminately kill untold numbers of owls, and housecats (feral and domestic) decimate birdlife (including owls) around the globe. These dangers and others have driven local and regional populations of some owl species into steep decline in many places, but at an individual and local level, they are problems we can each help solve.

Again, though, the big killer of entire species is habitat loss, including habitat fragmentation, loss of habitat to invasive species, and land-use changes on massive scales. Humans continue to alter the landscape, sometimes rapidly and extensively, and ecological diversity suffers apace. Certainly, deforestation of a remote tropical island that was once an idyllic paradise may seem beyond our scope of influence, but that's not always true. Many species, including numerous owl species, have their human advocates who work tirelessly to mitigate habitat loss and other perils. Worldwide, organizations and individuals continue to try to help the world's rare owls, and one way we can all help is to support these champions. At the same time, we can help owls avoid danger right in our own yards, properties, and neighborhoods by proactively addressing owl perils with oft-simple steps outlined here.

Ease the Speed

Collisions with vehicles send more owls to rehabilitation centers than any other cause, and those are just the injured owls—most collisions with vehicles are fatal. In southern Idaho, for example, some 1500 or more Barn Owls are killed on Interstate 84 every year; this 200-plus-mile stretch of freeway courses through prime Barn Owl habitat. And in Minnesota, the International Owl Center estimated that some 25 percent of the approximately 4000 Great Gray Owls that came to the state during the winter of 2004–05 were struck by vehicles. In fact, numerous researchers have concluded that collisions with vehicles may be the biggest source of mortality for owls because roadways offer the kind of

We can reduce owl deaths and injuries due to vehicle collisions by driving a little slower in owl habitat between dusk and dawn.

Owls are highly vulnerable to being struck by vehicles, not just because most species are active after dark but also because roadways provide good hunting grounds.

low-cover habitat ideal for many of the small mammals that owls eat. When hunting along roads, owls may be blinded by headlights or otherwise simply unable to avoid fast-moving vehicles. Moreover, some species, such as Barn Owls, typically fly within 3 meters of the ground when they hunt, placing them directly in harm's way.

In a study of Barred Owls in a suburban environment (around Charlotte, North Carolina), researchers discovered that both road design and driver behavior influenced owl mortality by vehicle collisions. Narrower roads seem to produce more collisions, perhaps because they tend to provide better habitat for Barred Owls in the form of readily available perches (roadside trees, signs, etc.) and narrower open areas to cross (Barred Owls are generally woodland owls that shun extensive open space). Driver behavior is largely about vehicle speed—both owls and drivers have less time to react to avoid collisions at higher vehicle speeds.

The authors of the Charlotte Barred Owl study urge further research but also offer straightforward ideas for reducing owl/vehicle collisions, concluding, "Our results indicate that the most effective measures to reduce Barred Owl road mortality are those based on driver behavior, i.e., vehicle speed, and roadway design, i.e., width. Given the unfeasibility of manipulating speed limits and road widths, we think the most practical approach to reducing the likelihood of Barred Owl–vehicle collisions in Charlotte, based on our results, is the use of owl crossing signs that exhort drivers to slow down. Such a measure was also suggested by Boves and Belthoff (2012) for the Barn Owl. . . . An awareness and education campaign in the neighborhoods where signs are placed should also be carried out over the long term."

During their study, these researchers (Sara A. Gagné, Jennifer Bates, and Richard Bierregaard) enjoyed excellent cooperation from residents who allowed them to install nest boxes and monitoring cameras on private property, and even conduct banding operations. Such amenability by homeowners, they realized, affirms that owls are fascinating and charismatic to many people, which might well predict success of an awareness and education campaign aimed at reducing owl/vehicle collisions.

These studies suggest that we can all do our part to reduce owl deaths and injuries caused by vehicle strikes. The simplest measure is to drive a little slower between dusk and dawn on roads that traverse owl habitat, whether the wide-open spaces where Barn Owls hunt or the forested uplands home to such species as the tiny Northern Saw-whet Owl.

I was 18 years old when I saw my first Northern Saw-whet Owl. Sadly, I killed it with my front windshield. As youngsters are apt to do, I was driving too fast, after dark, on a section of U.S. Highway 101 that climbs up and over the base of a headland on the Oregon coast. The collision was slight, but right away I was certain I'd hit an owl; I pulled over, walked back along the road, and found the crumpled tiny body. The experience haunted me, and forever after I tried to remember to ease off the pedal a bit wherever and whenever owls or any wildlife might be at heightened risk.

Beyond driving slower through owl habitat when owls are active, we can also look for opportunities to be proactive. If you live near, or routinely drive through, a place frequented by owls, investigate the possibility of erecting an owl crossing sign. Such signs are readily available but underutilized, and as residents in a Tulsa, Oklahoma, neighborhood discovered in 2016, sometimes all you need to do is ask. When a pair of Barred Owls decided to nest in the neighborhood a few years earlier, they faced a range of perils from humans, from being shot at to being struck by vehicles, so a few concerned residents formed an advocacy group called Barred Owls of Midtown Tulsa. When they realized the danger to the owls posed by traffic, the organizers appealed to the city, which responded within a week and even footed the bill for four "Warning, Low Flying Owls" signs.

Another way to reduce owl deaths by vehicle is to keep roadway margins clear of trash. Unfortunately, some people consider roads, especially quiet backcountry roads, to be their personal dump sites, a place to furtively unload garbage. And garbage can attract the small mammals that owls eat—mice, rats, and others—making such illicit trash piles into bait stations for owls. Signage can only do so much because people who secretly dump trash under cover of darkness already know they are acting illegally. But you can help mitigate the problem by forming (or joining an existing) Adopt-A-Highway or Sponsor-A-Highway program. These programs, active in many states and in Canadian provinces and territories, encourage volunteers to take charge of routine trash collecting on their sponsored section of roadway. Learn more by contacting your state or provincial department of transportation.

A LITTLE KNOWLEDGE GOES A LONG WAY

Red Owl (*Tyto soumagnei*)

RANGE: Madagascar
STATUS: Vulnerable, population decreasing (2016)
POPULATION: 2500–10,000

The Red Owl, aka Madagascar Red Owl, was known to science by 10 specimens collected between 1876 and 1934, and by a single sighting in 1973. Despite extensive zoological surveys on the island, that one sighting was the only evidence of the bird's existence for nearly 60 years. Then, in July 1993, naturalist Dominique Halleux was told of a captive Red Owl held by a resident of the village of Antanamangotroka, near Andapa in northeastern Madagascar. Halleux had been working on a World Wildlife Fund conservation project. When he examined the captive owl, he found it in poor health, and the bird's captor allowed Halleux to keep it.

Doing his best to nurse the owl back to health, Halleux relied on old reports about its ecology. *Bird Conservation International* (1994) detailed Halleux's efforts and discoveries: "Before the owl was transferred to D.H.'s care in late July 1993, it had been fed beef, frogs, rats and grasshoppers. Thereafter, on the basis of information in the literature concerning the Madagascar Red Owl's natural diet (Lavauden 1932), it was mostly given frogs—which it readily ate. It was fed twice per day, at dawn and dusk. The bird was only attracted to moving prey. In cases when the prey was dispatched by the owl and then dropped on to the floor of the cage, even within a few [centimeters] of the bird, the prey was not relocated. After catching frogs, the owl would hold the prey in its talons and crush the skull with its bill. After the frog was dead or immobile, the owl grasped it in its talons to crush the prey's bones. When the prey was large, greater than 6–7 cm, the owl would first remove and eat the head, and thereafter consume the rest of the body."

Upon taking possession of the owl, Halleux had noted that it appeared injured and possibly suffering partial paralysis due to a neurological disorder. Capture and a yearlong captivity had not been kind to the rare bird, and it succumbed that December. Thanks to Halleux's rediscovery and subsequent research, ornithologists have discovered that, though rare and seldom seen, Red Owls occupy a larger range than previously thought.

Part Ways with Poisons

Another leading cause of owl mortality—worldwide—is indirect (aka secondary) poisoning. Farmers, horticulturalists, orchardists, and even homeowners who use poison to control rodents such as mice and rats almost inevitably expose nontarget species to the same poison. Once a poison is in a rodent's system, it gets passed along to anything that eats that rodent. Death by rodent poison is especially acute with Barn Owls around the globe because these birds readily adapt to human trappings, including places where rodents congregate. But the problem goes much deeper, and studies have confirmed accumulated rodent poisons in a variety of owl species at frighteningly high frequencies. A 30-year study in Canada, for example, looked at raptors found dead or brought

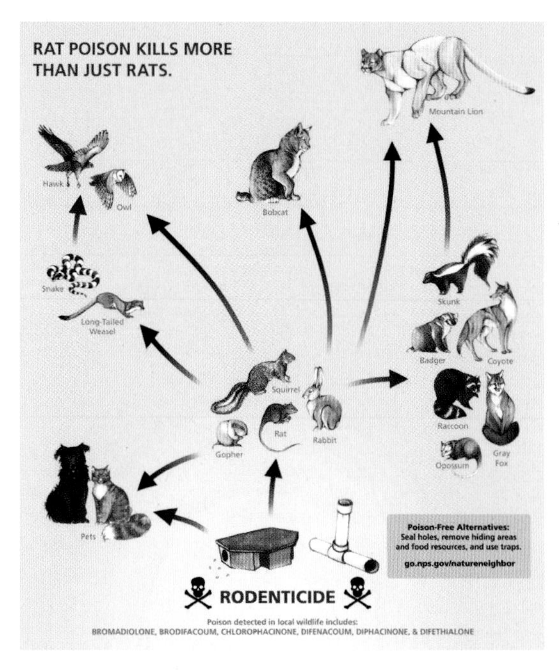

Pesticides, especially the various rodent poisons, are nefarious, frequently poisoning unintended victims such as owls.

into rehab centers: 96 percent of the Barred Owls and 81 percent of the Great Horned Owls in the study carried rodenticides in their systems.

Such findings, which are consistent across studies carried out worldwide, tell a woeful but cautionary tale: we need to quit relying on nondiscriminatory rodenticides. When fellow birding enthusiasts ask me what to do about burgeoning numbers of mice, rats, chipmunks, or squirrels attracted by backyard bird-feeding stations, I usually tell them what they don't want to hear: stop feeding the birds. At least for a little while.

Many of us enjoy watching birds that we draw into our yards with seed and suet, but the dark side to feeding wild birds is twofold: we attract unwanted species (like rodents) and we risk spreading disease among the birds themselves. Seldom do the birds need our charity, even during the harsh winters, so—because I enjoy watching birds at my birdfeeders—I mitigate the pitfalls with an on-again, off-again strategy. I provide seed and suet for a while, then back off for a week, then provide nothing for one to two weeks. During the downtime, I sterilize all my birdfeeders, suet cages, and shepherd's hooks to help keep a lid on avian diseases. Around the birdfeeders (and indoors) I sometimes contend with minor infestations of the common house mouse, so I use snap traps indoors and in outdoor spaces where I can place the traps with no risk to birds or other creatures. My birdfeeders also attract squirrels and chipmunks; they can be nuisances in some respects, but they come with the territory and tend to be dissuaded by the hiatuses in bird-feeding.

Those snap traps come in sizes large enough to dispatch average-size rats, and when sheltered so they can't kill or injure birds and other nontarget animals—including your pets—they are highly effective. Combined with cleanliness in all things related to bird feeding, they help substantially in controlling unwelcome rodents. Other nonpoisonous options include CO2 traps, such as the Goodnature A24. Never use glue traps, which trap indiscriminately and are notorious for snagging songbirds.

Rodent poisons are just one classification of pesticides. Many herbicides, insecticides, and other poisons can also easily end up in the food chain and ultimately kill nontarget species. Avoid these poisons when you can, and if you need pesticides, opt for formulas that target only the pest that is the issue.

Good Samaritans in Washington State rescued this Great Horned Owl that they found entangled in a soccer net.

Take Down the Nets

In my interviews with owl rehabilitators, I was surprised at how many of them mentioned seemingly innocuous soccer nets as inadvertent life-threatening traps for night-hunting owls. In fact, because they are active between dusk and dawn, owls are more prone to entanglement than any other species of raptor brought to rehab facilities. Net-entangled owls end up in rehab facilities all over the world. Even if their injuries are not life threatening, these birds can succumb to dehydration and exhaustion after struggling valiantly to free themselves; or they can succumb to capture myopathy, which results from the physical struggle and the stress from extreme fear, such as being captured and handled by humans.

Protecting owls from the threat of entanglement in soccer nets can be an individual, institutional, or community-wide endeavor: when and where possible, particularly where owls are known to live or where soccer fields occupy owl habitat, take down the nets at the end of the day. It's one of those simple solutions that takes nothing more than awareness of the problem and a willingness to help. Fake cobwebbing and spiderwebs, the kinds used for Halloween decorations, also snag unwitting owls at night, so use them indoors but not outdoors.

THE DANGEROUS GAME OF BIOLOGICAL CONTROLS

Laughing Owl (*Ninox albifacies*)

RANGE: Formerly New Zealand

STATUS: Extinct

Named for its exotic call, which one listener said was "a most unusual, weird cry which might almost be described as maniacal," the extinct Laughing Owl was a endemic to New Zealand. This unique owl spent a lot of time on and near the ground, and often hunted on foot, especially for insects, geckoes, and small birds (including kiwis). Its dietary preferences have been inferred from analysis of preserved and fossilized pellets, which accumulate at middens (basically piles of pellets at an owl's favorite roost).

The decline and ultimate extinction of this fascinating owl began when Polynesians first reached New Zealand centuries ago, inadvertently delivering stowaway Pacific rats to a rat-free paradise. The rats competed with the owls for prey and probably became owl nest raiders as well, with eggs making a protein-rich meal (in turn, Laughing Owls began preying on rats). To compound the owl's problems, European explorers first reached New Zealand in 1642 and their ships carried stowaway black rats. Wherever they took hold, invasive black rats were routinely devastating to endemic species.

Then came the rabbits; as standard trade fare among European seafarers, rabbits were brought to New Zealand around 1830. They were raised, they were released, and after a while, rabbit populations exploded. New Zealand simply had no predators capable of controlling such a fecund animal as the rabbit. So how did the Europeans deal with an explosive overabundance of feral rabbits? Simple: import predators that eat rabbits, specifically ferrets, domestic cats, and ermines (stoats). This early experiment in biological control failed. Rabbits bred so fast that the newcomer predators couldn't keep up; moreover, these predators found easy pickings in the native birds and mammals that had never seen a four-legged mammalian predator.

Laughing Owls (among many other endemic species) were doomed. Ironically, their fate was further sealed by the relentless zeal to collect specimens before the species was gone forever. The last Laughing Owl ever collected was found dead in 1914 (the same year the last living Passenger Pigeon died in the Cincinnati Zoo).

Be a Barbed Wire Warrior

Entanglement in barbed wire fences is a serious concern for many species of wildlife, but owls are especially vulnerable because of their nighttime hunting habits. There's no easy solution to the barbed wire menace. Quite the contrary, in fact: removing barbed wire is the answer, but it ain't easy. Barbed wire removal is laborious. It's just plain slow, hard work. I speak from firsthand experience: when barbed wire seriously injured one of my dogs, I asked the property owners if the old dilapidated three-strand fence served any purpose. The answer was no; they hadn't run livestock on that ranch in 30 years and I received permission to remove the wires if I could. With no way to drive to the fence, which ran for more than a mile just below the summit line of a steep ridge, I had to hike in with heavy-duty wire cutters, pliers, a pry bar, two pairs of leather gloves, and a backpack full of water and food. It took two days, but I removed all the wire, teaching myself how to do so in the process. I looped rolled-up hoops of barbed wire over remaining sturdy wooden fenceposts and my dogs have enjoyed safe, unfettered joy on that ridgeline ever since.

From time to time, I reflect on that project and wonder how many wildlife injuries I may have prevented and continue to prevent by the simple, though grueling, act of removing the barbed wire. The immediate area is home to Great Horned Owls, Western Screech-Owls, and Short-eared Owls. It was a drop in the bucket, considering that some 600,000 miles of barbed wire crisscross the American West, but increasingly, land managers, property owners, wildlife biologists, conservation organizations, and concerned individuals are working together to reduce the danger posed by barbed wire.

Barbed wire is so dangerous to wildlife that it should always be removed when it's not needed to contain livestock (and fences can be modified to make them safer for pronghorns and other animals). Even short sections of old fence can be lethal, so if you find a run-down or superfluous barbed wire fence, especially if it is within or near wildlife habitat, investigate the possibility of removing it. In fact, barbed wire removal on public and sometimes private lands (such as on properties administered by The Nature Conservancy and similar conservation organizations) frequently requires volunteer labor. Look to the Internet for such

PARADISE LOST

Bermuda Saw-whet Owl (*Aegolius gradyi*)

RANGE: Formerly resident of Bermuda
STATUS: Extinct

Science discovered the Bermuda Saw-whet Owl hundreds of years too late for any of us to ever see one. This tiny owl was mentioned scantly in the reports by English settlers who landed on Bermuda early in the 1600s but was not identified as a unique species until Storrs L. Olson (1944–2021), an ornithologist who spent his career at The Smithsonian, analyzed fossil owl bones collected from eight different sites on Bermuda.

Sadly, it seems this tiny owl was victimized by Bermuda's first human inhabitants—and the rats that inevitably infested ships of the European exploration and colonial era. The island, located in the Atlantic Ocean 600 miles off the United States, never had an indigenous human population. It was discovered by Spanish explorers in the early 1500s, then colonized a century later by English settlers. Prior to human arrival, Bermuda had no mammals other than bats, and its native fauna had no defense against invasive rats. Moreover, settlers inevitably deforested parts of the island of its native cedar and palmetto trees, denying the little owl the cavities it needed for nesting. But it seems the settlers had no use for the diminutive owl anyway. Captain John Smith (of Jamestown fame) reported on the English settlement as of 1623 and recorded that "there were a kinde of small Owles in great abundance, but they are now all slaine or fled."

As Olson wrote, "There has been no trace of any other kind of owl on Bermuda than *Aegolius gradyi*, so that we must assume that the owls seen by the early colonists were the same species that we find as fossils.

Olson surmised that the Bermuda Saw-whet Owl probably fed on birds and the "then-ubiquitous skink [a species of lizard], which was present on Bermuda for hundreds of thousands of years." Other potential foods include the extinct endemic Bermuda cicada, which probably went extinct because Bermuda was deforested of its native cedar trees by the accidental import of invasive juniper scale insects during the middle of the 20th century. Olson also realized that the little owl must have been diurnal, hunting largely during daylight when its prey species were active.

The Bermuda Saw-whet Owl was morphologically very similar to its probable progenitor, the Northern Saw-whet Owl (*A. acadicus*), and to this day, an occasional Northern Saw-whet Owl shows up on Bermuda during migration—rare individuals that stray far off course. Many thousands of years ago, a colonizing population of Saw-whet Owls would have been more proximal to Bermuda. Olson explains, "During the last glacial period, the breeding distribution of *A. acadicus* would have been depressed far to the south of its present range, putting most of the eastern populations closer to Bermuda throughout the year and thus probably facilitating colonization at a time when Bermuda was reaching its maximum land area during a period of lowered sea level."

possibilities and sign on to help, perhaps even as a family or peer-group project that will likely immerse you in beautiful environments.

For do-it-yourself barbed wire warriors, I can offer some instructions from experience. You will need heavy-duty wire cutters designed expressly to snip heavy-gauge metal wire; these are readily available at hardware and farm-and-ranch stores. You will also need small pry bars and pliers to remove nails and wire clips that attach wire to fenceposts, two pairs of heavy-duty leather gloves (one pair is a backup), eye protection (sunglasses or work goggles), and puncture-resistant clothing

Barbed wire is extremely hazardous to owls and other wildlife. The author removed more than a mile of old fence after he rescued an impaled Western Screech-Owl nearby.

(including long sleeves and work boots). You will roll barbed wire into large hoops, about 4 feet in diameter, and in my experience, you can roll two or even three strands together in the same hoop, but multistrand hoops get bulky and heavy very quickly, so two-person hoop crews are ideal.

Begin by removing wire from the fenceposts. After wire strands have been detached from a series of posts, cut the wires at one spot. Grasp the cut ends firmly and form a 4-foot-diameter hoop. Spiral the cut ends around the wire several times to secure the hoop. Now simply roll the hoop along the ground, hand over hand, gaining wire volume as you go. When the hoop becomes too large and heavy to wield, cut the wires to create new ends and wrap these ends firmly around the hoop several times to secure them. Repeat the process until the wire is gone. Where I dismantled more than a mile of fence, there was no way to transport the wire hoops away from the ridgeline, so I looped them over sturdy fenceposts at intervals, four or five hoops per post. Ideally, the wire should be removed and delivered to a metals recycling facility.

Mind the Mono

Discarded fishing line, often made of monofilament or fluorocarbon, which are both plastics that degrade slowly, pose a major threat of entanglement to all kinds of wildlife, including owls. It seems inconceivable that anglers would simply throw lengths and wads of fishing line on the ground with no regard for anti-littering laws or ethics, but such conduct is so routine that some states have initiated educational campaigns and installed special containers at popular fishing locations where people can dispose of fishing line. In addition to careless littering, fishing line also ends up discarded when anglers accidentally snag some obstruction they cannot reach, leaving no choice but to cut or break the line. But in my experience—I've been fly-fishing since I was a young boy—wanton littering accounts for most of the fishing line left behind by anglers. I've picked up and

Lost and discarded fishing line is a significant threat to owls and other animals. This Barred Owl was badly entangled.

Most owls rely on tree cavities and crevices for nesting and roosting.

discarded hundreds of yards of the stuff over the years, and given the scope of the problem, picking up the trash left by others seems to be the best remedy. So wherever your travels take you, whether close to home or far afield, always pick up any stray fishing line you find (beware of hooks possibly still attached to the end) and dispose of it in the nearest garbage can.

Leave the Trees

With few exceptions, owls need trees, and most small species of owls are cavity nesters that rely on woodpecker holes and natural cavities, which tend to be most numerous in dead and dying trees. So wherever possible (and safe) let the snags (upright dead trees) stand—they are prime habitat for woodpeckers, and thereby for nesting and roosting owls, and they often provide natural crevices as well, where limbs have broken off, where insects have bored into the wood, or where the tree has rotted. Live trees are also critical for owls. Many species of owls hide amid foliage or bows (limbs on coniferous trees, such as firs and pines) during the day; they find favorite hunting perches on tree branches, and their fledglings rely on horizontal limbs during the brancher stage of their young lives. So, if you need to prune trees on your property, wait until fall so as not to potentially disrupt nesting or fledging owls (and potentially many other bird species).

Don't Be a Neat Freak

Neatly manicured yards are lovely; I'm often intrigued and sometimes a little envious when I see a home surrounded by a beautifully kept yard, with hedges trimmed to perfection, flowers stair-stepped, gorgeous vines gracing trellises, and lawns cut as precisely as a golf course green. But such yards aren't ideal for wildlife, including owls. If you live in or near owl habitat, consider being a little less fastidious with your yard. Create a brush-pile as cover for songbirds and small mammals; maintain a "weed-lot" out in the open where low-growth grasses and other plants can grow to provide habitat for native voles and other potential owl prey; create borders, such as dense hedgerows, that provide cover for a variety of creatures. Install a water feature (owls may use it, but so will other birds) and make sure to keep the water clean.

At the same time, reduce hazards to owls. Screen off chimneys and ducts so small owls can't use them as roosts or even nests. Decorate windows with bird anti-collision stickers (widely available online and at nature stores) to prevent all sorts of birds, including owls, from being killed or injured by flying into reflective windows. Turn off outdoor lights before dusk, keep pets indoors between dusk and dawn, and reduce noise as best you can by limiting loud activities such as lawnmowing and weed-whacking, especially early in the morning and in the evening.

However, if you live near a busy road, busy railroad tracks, a busy industrial area, or in or near a place undergoing rapid development, it's best not to purposefully try to attract owls to your space, because doing so might endanger them. Always keep the best interest of the owls in mind.

Put Up an Owl Nest Box

Some owl species, especially screech-owls, are well known for their proclivity to use artificial nest boxes, but larger species, such as Barn Owls, Long-eared Owls, and Barred Owls will also use nest boxes that are properly located. Nest boxes specifically made for various species of owls are readily available, as are plans for making your own. Screech-owl nest boxes are generally placed at least 10 feet up in a tree, on a section of trunk free of branches, and in the quietest area of your property. Barred Owl boxes need to go higher, 15 to 30 feet above the ground, according to various sources. If you're uncomfortable on a tall ladder, hire an arborist or contact a local bird-rescue team or raptor center to ask if they work with a nest-box installer who can place the box for you.

Barn Owl boxes are best mounted on poles rather than trees. Barn Owls may also nest in a box located in an old barn or similar structure on your land, especially if people seldom visit the building. Platforms or nest

END OF TIME

Pernambuco Pygmy-Owl (*Glaucidium mooreorum*)

RANGE: Brazil
STATUS: Critically Endangered or Extinct
POPULATION: 0–50

Only after studying a single preserved specimen, and comparing it to similar local species, did Brazilian biogeographer José Maria Cardoso da Silva realize he was looking at a new species of owl. He and other scientists also realized the newly named Pernambuco Pygmy-Owl was critically endangered. The owl has been seen in only two locations within the Brazilian state of Pernambuco, where habitat loss from forest fragmentation threatens its existence, if indeed the species is still extant. This part of Brazil sits within the Atlantic Forest, one of the largest and most ecologically rich and diverse places on the planet, but the owl's tiny range is increasingly imperiled and already severely deforested. Near-complete destruction of the owl's habitat since the last sighting in 2001 led the IUCN to uplist the Penambuco Pygmy-Owl to Critically Endangered – Possibly Extinct in 2019.

Scientists involved in identifying this new but now perhaps extirpated species named it *mooreorum* after Intel founder Gordon Moore and his wife Betty; the couple's philanthropic organization, The Gordon and Betty Moore Foundation, is deeply involved in conservation. The Moores made one of the largest gifts in conservation history in 2001 with a 10-year-long series of grants totaling $261 million to Conservation International. The aim was to help the organization implement a major global strategy for biodiversity conservation. With luck, such foresight and philanthropy, coupled with dedicated scientists and conservationists, can save the Pernambuco Pygmy-Owl, but if it's already too late for this little owl, perhaps its story can inspire better stewardship for other critically endangered species.

cones installed in quiet, unused buildings or in trees can attract nesting Great Horned Owls, and Long-eared Owls also use nest baskets. Nest cones are structures that imitate large hawk nests, which are often commandeered by Great Horned Owls. They take a variety of forms, starting with a strong conical base made from wood, chicken wire, or even a big wicker basket. The inside of the cone is filled with sticks to mimic the typical nests made by Red-tailed Hawks and other raptors. Great Gray Owls adapt readily to nesting platforms made specifically for them.

In California, Barn Owl researcher Mark Browning discovered that installing nest boxes in Napa Valley vineyards plagued by abundant gophers led to a positive correlation between increasing Barn Owl populations and decreased gopher populations. Similar research projects by various private and public agencies and universities have yielded the same conclusions, demonstrating the value of Barn Owls (and other raptors) in controlling populations of crop-damaging rodents that can increase dramatically when farming usurps natural habitat needed to maintain appropriate densities of rodent predators. Farmers and landowners take note: encouraging owls and other raptors to reside on your property can be a highly effective part of pest-rodent mitigation. Excellent information is available from Western Sustainable Agriculture Research and Education (www.western.sare.org), Raptors Are The Solution (www.raptorsarethesolution.org), most raptor research and rehab centers, and many state wildlife agencies. Numerous nonprofits and businesses sell, and provide information about, owls boxes (see list of organizations at the end of this chapter).

In addition to nest boxes, add raptor perch poles to relatively open areas. Diurnal hunters—hawks and falcons—may use the poles during the day, and owls might find them equally useful after dark. Perch poles are ubiquitous in many agricultural settings but are also great for suburban settings. Place them at the edges of and, where possible, within fields, grasslands, meadows, and scrub-brush expanses. However, don't place them near roads or power lines. If you are handy with tools, perch poles are easy to make and install; otherwise, they are available from a variety of retail sources.

If you don't have a spot for an owl nest box or platform, consider volunteering to help build or install them elsewhere. Such activities are great fun—and highly educational—for school groups and other kid-centric organizations, but also for adults. To find out how you can get involved with owl nest box fun, check with state fish and wildlife agencies, owl rehab

Schoolkids in Alaska help with a U.S. Fish and Wildlife Service project to place nest boxes for Boreal Owls.

facilities, owl/wildlife advocacy groups, and with your nearest U.S. Fish and Wildlife Service, Bureau of Land Management, and U.S. Forest Service offices.

Watch the Cats and Dogs

Cats are preprogrammed to be voracious and effective predators, and those that spend a lot of time outdoors are especially treacherous to rodents and birds. Shockingly, domestic cats—housecats and feral housecats—kill some 2.4 billion birds annually just in the United States. They are the number one human-caused source of bird mortality. On top of that, cats kill more than 12 billion small mammals annually in the United States. The danger to owls is twofold: cats can easily kill small owl species and their young, and cats compete with native owls (and other raptors) for the small mammals that make up most of the diet of many birds of prey. Cat-injured owls routinely show up at rehab facilities, and during my research for *The Hummingbird Handbook* (2017) I learned that cats are such effective predators that they can easily swat speedy hummingbirds out of the air, ambushing the tiny birds by hiding beneath flowers and feeders.

I'm not anti-cat. I like housecats. I grew up with them and I have one. But cats need to be indoor pets for the sake of wildlife. A recent trend, as awareness of the prodigious lethality of housecats has spread, is to create outdoor environments for cats that keep them safely separated from wildlife. The common vernacular is "catio," a clever portmanteau for an outdoor cat habitat. They are widely available, and plans for building your own are ubiquitous on the Internet. I know it's hard to make the decision to reign in your free-roaming cat, but by providing a catio-style habitat with plenty of options for stimulation, you can have the best of both worlds—a well-adjusted, oft-stimulated pet and a local wildlife population that is a lot safer because of your conscientious action.

Dogs can also pose a threat to owls. Prey drive—the genetic urge to hunt for food—can lead dogs to attack and kill an owl that is on the ground, such as an adult owl that has

Housecats and feral housecats kill billions of birds and small mammals annually, posing a danger to small owls and impacting the prey supply for owls. Be very mindful about letting cats outdoors.

captured prey or a baby owl that has fallen or fluttered from a nest or branch. Be vigilant during the owl nesting and fledging season of winter and spring, and be mindful of where you run or walk your dog off-leash. Moreover, a well-trained dog is an easy dog, and I'm a big proponent of flawless recall: training your dog to come back to you without hesitation. I train my dogs to instantly return to me (recall) by voice, by whistle, and even by hand signal, and I do it all through positive-reinforcement training. That means that during the training paradigm, the dog gets a yummy treat for obeying. Over the years, I've been able to call my dogs away from wildlife on numerous occasions thanks to their obedience to recall.

Beware of Babies

Each spring, throughout the world, caring, concerned citizens find fledgling owls on the ground, assume they are abandoned by or separated from their parents, and deliver them to rehab centers. Often, these gangly little fuzzballs were not in any need of rescue. When they leave the nest, owls can't yet fly but they'll soon learn; juveniles of most species can climb and flutter and make their way about on branches using their feet, bills, and developing wings. Owls of this age are at their branching stage, when they leave the nest and take up short-term residence on a convenient limb near the nest. They can unceremoniously fall to the ground. All of this happens under the watchful eye of parents, but owls being owls—secretive and cryptic—the adults may not be obvious. Often the baby owl will climb back

Owl fledglings are usually best left alone, unless they are in imminent danger, such as in the case of this Barred Owl that ended up on a roadway.

into the tree, but if not, the parents typically continue caring for and feeding the fledgling.

Nestlings, on the other hand, are younger baby owls, not yet ready to venture beyond the protective confines of the nest. They are downy all over, with bare skin showing, unlike fledgling-age owls, which have begun growing flight feathers and tail feathers. If you find a nestling on the ground, return it to its nest if you can do so quickly and safely; otherwise, back away and call a local or regional rehab center.

If you find a fledgling owl on the ground, and see no obvious injuries, back away, distancing yourself from the bird. Then assess

FOR A FEW DOLLARS MORE

Sokoke Scops-Owl (*Otus ireneae*)

RANGE: Southeast Kenya and northeast Tanzania
STATUS: Endangered, population decreasing (2016)
POPULATION: 2,500–10,000

If modeled predictions hold true, the cute little Sokoke Scops-Owl, already endangered by loss of habitat on a massive scale, may face extinction, with the coup de grace issued by climate change. This owl's distribution in Africa is expected to shift southward and eastward in response to a warming climate by 2080, according to the authors of "Rapid decline and shift in the future distribution predicted for the endangered Sokoke Scops Owl *Otus ireneae* due to climate change," published in *Bird Conservation International* in 2012. "Very little of the future distribution of the Sokoke Scops Owl will overlap with current protected areas," according to the research paper, which demonstrates that this 6-inch-tall owl is in dire straits, especially because regions east and south of the owl's current range have very little appropriate habitat remaining and cannot recover these native habitats nearly fast enough even if humans had the will to attempt such a reversion.

A habitat-specific species, the Sokoke Scops-Owl survives in just two or three small pockets of forest in coastal Tanzania and adjacent Kenya. In Kenya's Arabuko-Sokoke Forest Reserve, where the owl was first discovered in 1965,

the bird only occurs in mixed *Brachylaena/Cynometra* forest despite the availability of other forest types. Such habitat-specific organisms are at heightened risk, unable to utilize other kinds of environments. The combination of widespread loss of these lowland forests and a rapidly changing climate may have doomed the Sokoke Scops-Owl, though the authors of the 2012 paper suggest a plan of action: translocate the owls to areas of suitable habitat not currently known to support the Sokoke Scops-Owl. Birds of the World (BOTW) authors list other proposed conservation measures: expand the Sokoke Forest Reserve, set aside 200 square kilometers of surrounding land to be managed for traditional uses only, and improve facilities and funding so forest officers can enforce conservation policies.

However, BOTW also notes the lack of funds to enforce conservation laws, which means illegal logging and forest clearing—the pursuit of the almighty dollar, so to speak—is difficult to stop. Data from The Peregrine Fund, which has sponsored much of the research into the Sokoke Scops-Owl, indicate the population declined by some 20 percent during a recent 15-year period. If it's too late to save this rare and secretive owl, its dark path to extinction illustrates the importance of financially supporting the work of organizations such as The Peregrine Fund that fund and undertake research and conservation work (see the list at the end of this chapter).

the situation. Is the fledgling in any immediate danger, such as from proximity to traffic or dogs or similar threats? If not, there's no need to take any action beyond calling your nearest wildlife rehab center. They can advise you over the phone or decide if an on-site visit is warranted. Take more immediate action only if the baby owl is imperiled or injured or if you can confirm that both parents have been killed, and the most important rule to follow if you believe the fledgling needs to be immediately rescued is to minimize human-to-fledgling contact. A baby owl can quickly imprint on a human, especially if you provide it food, and a fledgling that imprints on a human rather than its parents cannot be released back into the wild.

If the fledgling is in danger from traffic, move it away from the road and under the shelter of a tree or even onto a low branch. Ideally, a wildlife rehabber can then decide on the best course of action, but if you decide immediate danger requires removing the fledgling from the scene, be forewarned that a healthy fledgling may fight back with sharp claws and bill. If you have gloves available, wear them to protect yourself from the owl and from various parasites that can transfer to humans. Drape a towel, jacket, or sweatshirt over the owl to pick it up, then secure it in a box or other encloser. Keep it warm, do not offer food or water, do not interact with it, and get the bird to a rehab center as soon as possible. You've likely just saved an owl's life! By acting prudently and quickly, you have allowed rehabbers the ability to return the owl to the wild.

Invade the Invasives

Invasive blackberry is built to win, but its formidable defenses did not dissuade me from going to battle.

Where I live, in western Oregon, blackberry briars left unchecked quickly envelope everything. They overwhelm native plant communities, growing so prolifically that they shade out virtually everything else. Native to Eurasia, Himalayan Blackberry (*Rubus armeniacus*), also known as Armenian Blackberry, is ubiquitous in the Far West, especially west of the Sierra Nevada and Cascade Mountains from California into British Columbia, and is also well established in scattered locations elsewhere in the United States.

Armenian Blackberry bristles with rigid, needle-sharp thorns; it grows in dense tangles and snakes its way up into shrubs and trees; stems can grow 15 feet high and then arc earthward to creep along the ground or through other vegetation for dozens of feet. Like some alien menace from a 1950s sci-fi movie, these tendrils, upon reaching the ground, root from their tips. Dense thickets are impenetrable and so prolific that they easily overwhelm everything in their path, and not just shrubs and trees. Abandoned buildings and vehicles, deserted parking lots and roads—all are quickly subjugated by these massively armed invaders that form extensive roots systems, further imperiling native plants by usurping available water.

In the half-mile-long strip of native hardwood forest near my home, the blackberries had all but declared victory. But when the farmer who owned that small, forested tract gave me permission to cut a path through it to run my dog, I had to clear enough of the old dirt-track access road to create a passable

Battling invasive plants that overwhelm landscapes is no easy task, but clearing non-native blackberry and similar sun-suckers helps restore native ecosystems, in this case low-growth understory ideal for rodents and the owls that hunt them.

trail. That winter, by machete and then gas-powered weedwhacker, I opened the trail for 400 yards, much to the delight of my Weimaraner, Java. And back in the woods near the end of the trail, I discovered a large nest high in an oak tree—the work of Red-tailed Hawks, but apparently last used by Great Horned Owls judging by a few scattered feathers on the ground below.

Often, during the winters, Great Horned Owls hooted in the night, and now I'd discovered their nest site. Or had I? The few feathers I found were badly degraded; they'd been there a while. I watched the nest all winter and into spring but never saw owls. Spring arrived in earnest, and in parts of the pathway recently liberated from dense blackberry brambles, native wildflowers—*Trillium*, waterleaf, and

SOME INVASIVE SPECIES SAIL SHIPS

Siau Scops-Owl (*Otus siaoensis*)

RANGE: Siau Island (Indonesia)
STATUS: Critically Endangered or Extinct (2018)
POPULATION: 0–49

The ages-old story of invasive species and deforestation have probably exterminated the last of the Siau Scops-Owls, a species known from just a single specimen collected in 1866 by Dutchman Renesse van Duivenbode. Taxonomy of this small owl—the one known specimen measures just 7.5 inches long—isn't mutually agreed upon, as some authors consider it a subspecies of the Sulawesi Scops-Owl or the Moluccan Scops-Owl, but the International Ornithologists' Union recognizes it as a distinct species. Even before Dutch colonials converted much of the island's forest to plantations for coconuts, nutmeg, and cloves, the population of Siau Scops-Owls was likely quite small; the island is only 10 miles long and its upper volcanic slopes lack dense vegetation.

According to Cornell University's Birds of the World, a 32-day survey in 2006, along with subsequent searches in 2009 and 2016, failed to find the species. The authors of the Birds of the World account say, "Further searches are needed to determine whether the species is still extant and it has been suggested that a tiny, unsurveyed remnant of primary forest on N slope of Karangetang volcano, which is probably accessible only by boat, offers the best chance of rediscovering the species."

Island endemics—those species that evolved on isolated islands—frequently occupy very limited ranges and tend to be habitat specific. Birds, including many owls, have suffered tremendously from introduction of non-native animals (and plants), with numerous species having gone extinct within the past few hundred years. Three species of rats, which inevitably lived aboard seafaring vessels and are non-native to most of the world's islands, began spreading with humanity; Polynesian explorers spread rats to Pacific islands even before European explorers delivered the nefarious mammals to islands all around the globe. Rats eat eggs, devour food native species rely on, and breed rampantly. They aren't the only damaging invasive—others include pigs and other hooved mammals, cats, dogs, mice, rabbits, a host of small predators used to try to control invasive rats, and others. Most islands had no mammals prior to the arrival of man, so birds and other life evolved no defenses against predation by, competition from, and diseases spread by the invaders.

In 2012, Indonesia released this set of postage stamps depicting some of the nation's endangered and threatened birds, including the Siau Scops-Owl (upper left).

Given the thousand-odd-year history of man's exploration of oceanic islands, we have no idea how many owls (and other birds) went extinct before their presence was ever recorded. For some species, only a fossil record remains. On Hawaii, for example, the arrival of Polynesians a thousand years ago quickly spelled doom for four species of extinct Stilt-Owls, representing a genus that no longer exists. Like the Siau Scops-Owl, many rare species of island-endemic owls today face potential if not inevitable extinction from the anthropogenic triumvirate of rampant deforestation, invasive species, and climate change.

camas—grew and bloomed. Their root systems had apparently survived the invasion of the blackberries and all they needed to begin sprouting and blooming again was sunlight.

Inspired by the liberated flowers, I went on the attack and experimentally cleared away blackberry from a gentle sloping 50-by-20-yard plot below the trail. The impact was remarkable: native plants flourished. A multispecies low-growth understory replaced the 8-foot-high impenetrable blackberry thicket. But keeping the blackberries at bay required regular machete work, digging out many of the corms (root balls), and later in the year applying targeted herbicide sparingly. I expanded the blackberry-free zone over the next two years and soon had created a parklike setting amid the oaks, ashes, and other native trees.

Many owls brought to rescue and rehab facilities cannot be released back into the wild. Some facilities provide the public the chance to symbolically adopt such owls with donations that support the bird's care.

And then came the owls. Great Horned Owls used the old hawk nest one early spring; the ensuing summer brought Barred Owls to the little forest—the first I'd ever seen since living there. Did significant blackberry removal lead directly to the presence of owls? I don't know. But with that cleared land came a thriving complex ecological community, which replaced what amounted to monoculture, a blackberry-dominated landscape that repressed species diversity. I'd like to think that the restored low-growth understory created ideal habitat for native voles, shrews, chipmunks, and pocket-gophers, which in turn created hunting habitat for owls.

The blackberry-free zone down in the woods continues to provide surprises, but maintaining it requires constant work. The Great Horned Owls have returned each winter; Barred Owls prowl the woods in the summer. And one spring night we were delighted to hear the unmistakable call of a Western Screech-Owl echoing eerily back in the woods. Perhaps the presence of the owls is coincidental to my war on invasive blackberries; however, though my evidence to the contrary is purely empirical, I think the improved habitat is causative. Moreover, my little experimental parkland now attracts a wider array of birds in general than it did prior to the blackberry removal. The project continues to be an abject lesson in the value of maintaining complex native ecosystems even at the price of substantial sweat equity.

And above all else, owls need intact habitat and complex ecosystems. Invasives such as blackberry—and notorious English ivy, among many others—can overwhelm the landscape and repress complexity at the expense of many native species. The solution is to reduce invasives in favor of native species. It takes work, but opportunities abound, from battling noxious invasives in your own space to volunteering to help remove blackberries, ivy, and other non-native species at parks and other public places.

Adopt an Owl

Even if you live in a thoroughly urban setting completely lacking in owls and owl habitat, you can help these enigmatic birds through owl adoption programs. No, you won't be tasked with housing and caring for an owl, but by donating a modest sum, you can support critical owl research or help rehab facilities offset the costs of rescuing, treating, and housing sick and injured owls. With adopt-an-owl programs run by organizations such as the Owl Research Institute and Whitefish Point Bird Observatory, for example, you symbolically adopt an owl and contribute directly to research that is critical to conservation efforts. Groups like the Alaska Wildlife Conservation Center and The Owls Trust (Wales), on the other hand, allow you to adopt an owl in their care by making one-time or recurring donations. These are just a few of the many great organizations that offer adopt-an-owl programs.

This Eastern Screech-Owl injured itself when it collided with a window in the dark. Wildlife veterinarians and rehabbers help thousands of injured owls annually.

Support the Owl Peeps

Adopt-an-owl programs are ubiquitous and important, but you can also help researchers and rehabilitators fund their critical work with donations of all sizes, and you can find many opportunities to do volunteer work to benefit owls. You can start at the local level by searching for rehab facilities in your area (try searching online for "raptor advocacy groups" or "raptor centers"). Many such facilities fly under the radar, with small staffs helping injured owls and other wildlife; some combine rescue and rehab work. In gathering data for this book, I interviewed rehab center personnel from around the world, and without exception, their enthusiasm is infectious and their passion is inspiring. In addition, various nonprofit research and advocacy organizations specialize in owls and other raptors, and they also need donations to carry on their important, seminal work. These groups, through their ongoing research and advocacy for owls, are vital to the science and understanding of these enigmatic birds.

Among the numerous owl advocacy, research, and conservation organizations found throughout the world, the International Owl Center based in Minnesota is unique. It is the

LONG WAIT FOR A LITTLE OWL

Santa Marta Screech-Owl (*Megascops gilesi*)

RANGE: Santa Marta Mountains of northeast Colombia
STATUS: Vulnerable, population decreasing (2019)
POPULATION: 2,300–7,500

In 1919, American ornithologist Melbourne Armstrong Carriker was convinced he'd collected a new species of screech-owl in northernmost Colombia and sent the sample to his colleague W. E. Clyde Todd at the Carnegie Museum in Pittsburgh, Pennsylvania. Lacking sufficient data to firmly identify the specimen as a new species, Todd and Carriker assigned it as a subspecies of the Tropical Screech-Owl (*Megascops choliba*) in 1922. The single specimen was largely forgotten even though the bird itself was seen regularly in its diminutive home range. Decades later in the 1990s, one of the world's most prolific bird call/song recording experts, Peter Boesman, along with bird vocalization expert Paul Coopmans, recorded the owl's unique calls, but their tapes remained unpublished. Then in 2007, Danish ornithologist Niels Krabbe was also able to record the owl's calls and finally authored a scientific description of the little owl, which was published in 2017. The Santa Marta Screech-owl was officially recognized as a species.

Carriker, who passed away in 1965 at age 85, traveled extensively in South America and collected specimens for the Carnegie Museum, the American Museum of Natural History, the Field Museum of Natural History, and others. He was assistant curator of birds at Carnegie Museum from 1907 to 1909, but settled in Santa Marta in 1911 and began extensive bird collecting. Todd (1874–1969) hired on as an assistant at the Carnegie Museum in 1898 and spent the rest of his career there. He made more than 20 expeditions to the arctic, leading to his *Birds of the Labrador Peninsula and Adjacent Areas* (1963), but years earlier, in 1922, he and Carriker coauthored *The Birds of the Santa Marta Region of Colombia: A Study in Altitudinal Distribution*. A bout with malaria, which he contracted in Washington, DC, prevented Todd from ever visiting Central or South America, but he diligently studied Carriker's specimens and field notes and the two produced a valuable account of avian diversity in this small, ecologically rich corner of Colombia.

Todd was ahead of his time as a conservationist, voicing concern over habitat fragmentation and even global warming, not to mention his objection to excessive collecting of birds and their eggs by museums. In the 1940s, he transferred the 71-acre parcel where his grandfather had farmed to the Audubon Society of Western Pennsylvania on the condition that the property be made into a nature preserve; in 1965, Todd donated an adjacent 61 acres. Today, the Todd Nature Reserve totals 334 acres and visitors have recorded some 180 bird species at the site—including Barred Owls, Great Horned Owls, Eastern Screech-Owls, and even Northern Saw-whet Owls.

nation's only all-owl education center, and each March hosts a unique multiday International Festival of Owls, which includes a bedazzling slate of interactive activities for both kids and adults, along with expert speakers, live owl programs, owl prowls, and much more. For owl fans, this event is the apogee, and well worth a springtime weekend in the small southeastern Minnesota town of Houston (about a two-hour drive from Minneapolis–Saint Paul). The Center is open Friday through Monday year-round, with a variety of educational programs held each day, along with other special annual events.

5

Owls of the World

An Annotated Gallery

Taxonomy of owls is challenging, especially with species and potential subspecies for which the record is scant. Many owls are island endemics, meaning they are native to one or more islands and through genetic isolation have evolved slightly different morphological characteristics than parent populations that originally colonized their range. Similarly, owls can experience extensive speciation in ecologically rich regions that offer advantages to creatures that can evolve to fill specific niches. That's why bird diversity is highest in fertile equatorial forests: Central and South America, for example, are home to 20 or so species of screech-owls alone, not to mention dozens of other owls.

Classifying owls—assigning them to a genus, and in many cases deciding whether they are a species or subspecies—is dynamic science. Changes to taxonomy are routine as scientists try to unravel

Mottled Owl (*Strix virgata*), photographed in Campeche, Mexico genetic relationships and evolutionary lineages. For rare, seldom seen (or heard), and extinct owls, biological evidence can be scant. Taxonomists rely on museum specimens (many of them collected in the 19th century), sightings and photographs, historical and contemporary accounts by natives and explorers, recordings of vocalizations, fossil evidence, faunal remains (bones, regurgitated pellets, feathers), genetic evaluation, and more to learn about owls and try to classify them. Sometimes molecular evidence reveals that very similar-looking owls are not that closely related, and sometimes divergent-looking owls turn out to be closely akin to one another. Owl taxonomy is also fascinating because sometimes entirely new species come to light via reexamination of old specimens and old data.

This chapter features photos of some of the world's nearly 250 owl species, a number that is subject to change as scientists continue to study these enigmatic raptors (the remaining species are listed at the end of this chapter), and also indicates their range and their relative abundance, or status, which (unless otherwise noted) is gleaned from the International Union for Conservation of Nature Red List of Threatened Species (www.iucnredlist.org), the world's most comprehensive inventory of the conservation status of species from around the world. For endangered owls, the estimated remaining population is also listed. Some owls may have relatively robust and stable populations in general but subspecies that are in danger of extinction, or that are already extinct. For example, the Burrowing Owl is widespread in the Americas, yet recent evidence has demonstrated that they were extirpated from two islands in the Lesser Antilles within the historic period, and Burrowing Owls are endangered in Canada and threatened in Mexico.

Numerous owl species and subspecies face potential extinction, but luckily, as a group, owls have many advocates. These wide-eyed, mysterious birds elicit awe and curiosity, their intrigue evident in the photos that follow, showing many of the world's fascinating owls.

African Barred Owlet (*Glaucidium capense*)
Range: Southern Africa
Status: Least Concern, population decreasing (2016)

African Grass-Owl (*Tyto capensis*)

Range: Sub-Saharan (central and southern) Africa
Status: Least Concern, population decreasing (2016)

African Scops-Owl (*Otus senegalensis*)

Range: Africa
Status: Least Concern, population stable (2016)

African Wood-Owl (*Strix woodfordii*)

Range: Africa
Status: Least Concern, population stable (2016)

Andean Pygmy-Owl (*Glaucidium jardinii*)

Range: Venezuela and Peru
Status: Least Concern, population stable (2016)

Asian Barred Owlet (*Glaucidium cuculoides*)

Range: Himalayas to Southeast Asia
Status: Least Concern, population increasing (2016)

Austral Pygmy-Owl (*Glaucidium nana*)

Range: Southern Cone of Argentina and Chile
Status: Least Concern, population stable (2016)

Australian Masked-Owl (*Tyto novaehollandiae*)

Range: Australia and south-central Papua New Guinea
Status: Least Concern, Population stable (2016)

Bare-shanked Screech-Owl (*Megascops clarkii*)

Range: Middle and South America
Status: Least Concern, population stable (2022)

Barking Owl (*Ninox connivens*)

Range: Moluccas, Papua New Guinea, Australia
Status: Least Concern, population decreasing (2016)

Black-and-white Owl (*Strix nigrolineata*)

Range: Mexico to Venezuela and Peru
Status: Least Concern, population decreasing (2018)

"Eastern" Barn Owl (*T. alba javanica*)

Barn Owl (*Tyto alba*)

Range: Extensive range in the Americas, Europe, Africa, the Middle East, Southeast Asia, southwest Pacific islands, Australia. Previously, *Tyto alba* was differentiated from "American Barn Owl" and "Eastern Barn Owl," and the taxonomy seems still unresolved, but the American Ornithological Society abides by the single-species classification.
Status: Least Concern, population stable

Black-banded Owl (*Strix huhula*)
Range: South America
Status: Least Concern, population decreasing (2016)

Black-capped Screech-Owl (*Megascops atricapilla*)
Range: Southeast Brazil/eastern Paraguay
Status: Least Concern, population stable (2016)

Brown Boobook (*Ninox scutulata*)
Range: India, South-central Asia, Southeast Asia
Status: Least Concern, population stable (2021)

Cape Eagle-Owl (*Bubo capensis*)
Range: Africa
Status: Least Concern, population stable (2016)

Chaco Owl (*Strix chacoensis*)

Range: Argentina, southern Bolivia, western Paraguay
Status: Near Threatened, population decreasing
Population: Unknown

Chestnut-backed Owlet (*Glaucidium castanotum*)

Range: Sri Lanka
Status: Near Threatened, population decreasing (2016)
Population: 10,000–20,000

Cinnamon Screech-Owl (*Megascops petersoni*)

Range: Andes Mountains, Ecuador, and northwest Peru
Status: Least Concern, population decreasing (2016)

Colima Pygmy-Owl (*Glaucidium palmarum*)

Range: Western Mexico
Status: Least Concern, population decreasing (2020)

Collared Owlet (*Taenioptynx brodiei*)

Range: Central and Southeast Asia
Status: Least Concern, population decreasing (2021)

Collared Scops-Owl (*Otus lettia*)

Range: Himalayas, eastern Asia, Southeast Asia
Status: Least Concern, population stable (2016)

Crested Owl (*Lophostrix cristata*)

Range: Middle and South America
Status: Least Concern, population decreasing (2018)

Eurasian Eagle-Owl (*Bubo bubo*)

Range: Widespread in Eurasia
Status: Least Concern, population decreasing (2016)

Eurasian Pygmy-Owl (*Glaucidium passerinum*)

Range: Eurasia
Status: Least Concern, population stable (2016)

Foothill Screech-Owl (*Megascops roraimae*)

Range: Northern South America
Status: Least Concern (according to Birds of the World, Cornell University)

Forest Owlet (*Athene blewitti*)

Range: India
Status: Endangered, population decreasing (2018)
Population: 250–1000

Giant Scops-Owl (*Otus gurneyi*)

Range: Samar, Dinagat, Siargao, and Mindanao (Philippines)
Status: Vulnerable, population decreasing (2016)
Population: 2500–10,000

Himalayan Owl (*Strix nivicolum*)
Range: Himalayas to eastern Asia
Status: Least Concern, population stable (2016)

Jungle Owlet (*Glaucidium radiatum*)
Range: India
Status: Least Concern, population stable (2016)

Koepcke's Screech-Owl (*Megascops koepckeae*)
Range: Peru
Status: Least Concern, population stable (2016)

Least Pygmy-Owl (aka East Brazilian Pygmy Owl, aka Brazilian Pygmy-Owl) (*Glaucidium minutissimum*)
Range: Brazil and Paraguay
Status: Least Concern, population decreasing (2018)

Lesser Horned Owl (aka Magellanic Horned Owl) (*Bubo magellanicus*)

Range: South America
Status: Least Concern, population stable (2016)

Little Owl (*Athene noctua*)

Range: Europe, North Africa, Asia
Status: Least Concern, population stable (2018)

Long-whiskered Owlet (*Xenoglaux loweryi*)

Peru
Status: Vulnerable, population stable (2020)
Population: 250–1000

Madagascar Owl (*Asio madagascariensis*)

Range: Madagascar

Status: Least Concern, population decreasing (2018)

Madagascar Scops-Owl (aka Rainforest Scops-Owl) (*Otus rutilus*)

Range: Madagascar
Status: Least Concern, population stable (2016)

Mantanani Scops-Owl (*Otus mantananensis*)

Range: Philippines
Status: Near Threatened, population decreasing (2016)
Population: 6000–15,000

Morepork (*Ninox novaeseelandiae*)

Range: New Zealand and Norfolk Island
Status: Least Concern, population stable (2016)

Mottled Owl (*Strix virgata*)

Range: Middle and South America
Status: Least Concern, population decreasing (2022)

Northern White-faced Owl (*Ptilopsis leucotis*)

Range: Central Africa
Status: Least Concern, population stable (2016)

Ochre-bellied Boobook (*Ninox ochracea*)

Range: Buton, Peleng, and Sulawesi Islands (Indonesia)
Status: Near Threatened, population decreasing (2016)
Population: 10,000–20,000

Oriental Bay-Owl (*Phodilus badius*)

Range: South-central Asia through Southeast Asia
Status: Least Concern, population stable (2016)

Oriental Scops-Owl (*Otus sunia*)

Range: Eastern Asia (including Japan), Southeast Asia, Himalayan Asia, India
Status: Least Concern, population stable (2021)

Pacific Screech-Owl (*Megascops cooperi*)

Range: Southern Mexico to Costa Rica
Status: Least Concern, population decreasing (2020)

Palawan Scops-Owl (*Otus fuliginosus*)

Range: Palawan (Philippines)
Status: Near Threatened, population decreasing (2016)
Population: 10,000–20,000

Pearl-spotted Owlet (*Glaucidium perlatum*)

Range: Sub-Saharan Africa
Status: Least Concern, population stable (2016)

Powerful Owl (*Ninox strenua*)

Range: Australia
Status: Least Concern, population stable (2016)

Rufescent Screech-Owl (*Megascops ingens*)
Range: Andes Mountains, Venezuela to Bolivia
Status: Least Concern, population decreasing (2016)

Rufous Owl (*Ninox rufa*)
Range: Papua New Guinea and northern Australia
Status: Least Concern, population decreasing (2016)

Rufous-banded Owl (*Strix albitarsis*)
Range: Venezuela to Bolivia
Status: Least Concern, population decreasing (2018)

Rufous-legged Owl (*Strix rufipes*)
Range: Central Chile and adjacent Argentina through Tierra del Fuego
Status: Least Concern, population decreasing (2016)

Sooty Owl (*Tyto tenebricosa*)

Range: New Guinea and southeast Australia
Status: Least Concern, population decreasing (2016)

Southern Boobook (*Ninox boobook*)

Range: Australia, Sumba (Indonesia), Timor, Papua New Guinea
Status: Least Concern, population stable (2016)

Southern White-faced Owl (*Ptilopsis granti*)

Range: Southern Africa
Status: Least Concern, population stable (2016)

Spectacled Owl (*Pulsatrix perspicillata*)

Range: Middle and South America
Status: Least Concern, population decreasing (2018)

Spotted Eagle-Owl (*Bubo africanus*)

Range: Africa

Status: Least Concern, population stable (2022)

Spotted Owlet (*Athene brama*)

Range: India and Southeast Asia

Status: Least Concern, population stable (2016)

Spotted Wood-Owl (*Strix seloputo*)

Range: Southeast Asia

Status: Least Concern, population stable (2016)

Striped Owl (*Asio clamator*)

Range: South and Middle America

Status: Least Concern, population decreasing (2020)

Stygian Owl (*Asio stygius*)

Range: South and Middle America, Caribbean Islands
Status: Least Concern, population decreasing (2016)

Sulawesi Scops-Owl (*Otus manadensis*)

Range: Sulawesi and adjacent islands (Indonesia)
Status: Least Concern, population decreasing (2022)

Sunda Scops-Owl (*Otus lempiji*)

Range: Thailand, Malay Peninsula, Sumatra, Borneo, Java
Status: Least Concern, population stable (2016)

Tanimbar Boobook (*Ninox forbesi*)

Range: Tanimbar Islands (Indonesia)
Status: Least Concern, population decreasing (2016)

Tawny Owl (*Strix aluco*)

Range: Western Eurasia
Status: Least Concern, population stable (2016)

Tawny-bellied Screech-Owl (*Megascops watsonii*)

Range: Amazonian South America
Status: Least Concern, population decreasing (2018)

Tropical Screech-Owl (*Megascops choliba*)

Range: Middle and South America
Status: Least Concern, population decreasing (2020)

Unspotted Saw-whet Owl (*Aegolius ridgwayi*)

Range: Central America to Chiapas, Mexico
Status: Least Concern, population stable (2016)

Ural Owl (*Strix uralensis*)

Range: Widespread through northern Eurasia; southeast and central Europe

Status: Least Concern, population stable (2021)

White-browed Owl (*Athene superciliaris*)

Range: Madagascar

Status: Least Concern, population decreasing (2016)

White-throated Screech-Owl (*Megascops albogularis*)

Range: South America

Status: Least Concern, population stable (2016)

Yungas Pygmy-Owl (*Glaucidium bolivianum*)

Range: Peru and Argentina

Status: Least Concern, population decreasing (2016)

Yungas Screech-Owl (aka Montane Forest Screech-Owl) (*Megascops hoyi*)
Range: Southern Bolivia to northwest Argentina
Status: Least Concern, population decreasing (2016)

The Rest of the World's Owls

Abyssinian Owl (aka African Long-eared Owl) (*Asio abyssinicus*)

Range: Albertine Rift (Democratic Republic of Congo, Rawanda, Uganda), Ethiopia, Kenya
Status: Least Concern, population stable (2016)

Akun Eagle-Owl (*Ketupa leucostictus*)

Range: Central and western Africa
Status: Least Concern, population decreasing (2018)

Albertine Owlet (*Glaucidium albertinum*)

Range: Albertine Rift of northeast Democratic Republic of Congo and Rwanda
Status: Near Threatened, population decreasing (2021)
Population: 2500–10,000

Alor Boobook (*Ninox plesseni*)

Range: Alor and Pantar Islands (Indonesia)
Status: Unknown; reclassified as a separate species from Southern Boobook in 2019.

Amazonian Pygmy-Owl (*Glaucidium hardyi*)

Range: Amazonian South America
Status: Least Concern, population decreasing (2019)

Andaman Boobook (*Ninox affinis*)

Range: Andaman Islands (India)
Status: Least Concern, Population decreasing (2017)

Andaman Masked-Owl (*Tyto deroepstorffi*)

Range: Andaman Island (India)
Status: Taxonomy under revision

Andaman Scops-Owl (*Otus balli*)

Range: Andaman Island (India)
Status: Least Concern, population stable (2017)

Anjouan Scops-Owl (*Otus capnodes*)

Range: Anjouan Island (Africa)
Status: Endangered, population decreasing (2017)
Population: 2300–3600

Annobón Scops-Owl (*Otus feae*)

Range: Annobón Island (Republic of Equatorial Guinea)
Status: Critically Endangered, population decreasing (2016)
Population: 50–250

Arabian Eagle-Owl (*Bubo milesi*)

Range: Southern Arabian Peninsula
Status: Least Concern, population stable (2022)

Arabian Scops-Owl (*Otus pamelae*)

Range: Saudi Arabia, Yemen, Oman
Status: Least Concern, population stable (2016)

Ashy-faced Owl (*Tyto glaucops*)

Range: Hispaniola (Caribbean)
Status: Least Concern, population stable (2016)

Baja Pygmy-Owl (*Glaucidium hoskinsii*)

Range: Baja California (Mexico)
Status: Least Concern, population trend unknown (2016)

Balsas Screech-Owl (*Megascops seductus*)

Range: Southwest Mexico
Status: Least Concern, population decreasing (2020)

Band-bellied Owl (*Pulsatrix melanota*)

Range: Colombia to Bolivia
Status: Least Concern, population decreasing (2018)

Banggai Scops-Owl (*Otus mendeni*)

Range: Banggai Island (Indonesia)
Status: Vulnerable, population decreasing (2016)
Population: 2500–10,000

Bare-legged Owl (*Margarobyas lawrencii*)

Range: Cuba
Status: Least Concern, population stable (2016)

Barred Eagle-Owl (*Ketupa sumatranus*)

Range: Malay Peninsula and Greater Sundas Islands
Status: Near Threatened, population decreasing (2021)
Population: Unknown

Barred Owl (*Strix varia*)

Range: North America
Status: Least Concern, population increasing (2016)

Bearded Screech-Owl (*Megascops barbarus*)

Range: Middle America
Status: Least Concern, population decreasing (2020)

Bismark Boobook (*Ninox variegata*)

Range: Lavongai and New Ireland Islands (Papua New Guinea)
Status: Least Concern, population decreasing (2018)

Biak Scops-Owl (*Otus beccarii*)

Range: Biak and Supiori Islands (Indonesia)
Status: Vulnerable, population decreasing (2016)
Population: 2500–10,000

Blakiston's Fish-Owl (*Ketupa blakistoni*)

Range: Siberia, northeastern China, Korean Peninsula, Japan
Status: Endangered, population decreasing (2016)
Population: 1000–2500

Boreal Owl (*Aegolius funereus*)

Range: Circumpolar at northern latitiudes
Status: Least Concern, population stable (2021)

Brown Fish-Owl (*Ketupa zeylonensis*)

Range: Middle East to Southeast Asia
Status: Least Concern, population decreasing (2016)

Brown Wood-Owl (*Strix leptogrammica*)

Range: Himalayas, India, China, Southeast Asia
Status: Least Concern, population decreasing (2016)

Buff-fronted Owl (*Aegolius harrisii*)

Range: South America
Status: Least Concern, population stable (2016)

Buffy Fish-Owl (*Ketupa ketupu*)

Range: Southeast Asia
Status: Least Concern, population stable (2016)

Burrowing Owl (*Athene cunicularia*)

Range: North, Central, and South America
Status: Least Concern, population decreasing (2016)

Buru Boobook (*Ninox hantu*)

Range: Buru (Indonesia)
Status: Least Concern, population decreasing (2016)

Camiguin Boobook (*Ninox leventisi*)

Range: Camiguin Island (Philippines)
Status: Vulnerable, population stable (2021)
Population: 250–1400

Cebu Boobook (*Ninox rumseyi*)

Range: Cebu Island (Philippines)
Status: Vulnerable, population stable (2021)
Population: 250–1000

Central American Pygmy-Owl (*Glaucidium griseiceps*)

Range: Middle America
Status: Least Concern, population stable (2016)

Choco Screech-Owl (*Megascops centralis*)

Range: Panama, Columbia, Ecuador
Status: Not yet evaluated (according to Birds of the World, Cornell University)

Chocolate Boobook (*Ninox randi*)

Range: Philippines (except Palawan group) and Talaud Island (Indonesia)
Status: Near Threatened, population decreasing (2016)
Population: Unknown

Christmas Island Boobook (*Ninox natalis*)

Range: Christmas Island (territory of Australia in the Indian Ocean)
Status: Vulnerable, population stable (2022)
Population: 350 (estimates from various surveys range from 240 to 1200)

Cinereous Owl (*Strix sartorii*)

Range: Mexico
Status: Unknown but probably very rare and decreasing

Cinnabar Boobook (*Ninox ios*)

Range: Sulawesi Island (Indonesia)
Status: Least Concern, population stable (2021)

Cloud-forest Pygmy-Owl (*Glaucidium nubicola*)

Range: Colombia and Ecuador
Status: Vulnerable, population decreasing (2016)
Population: 1500–7000

Cloud-forest Screech-Owl (*Megascops marshalli*)

Range: Peru
Status: Near Threatened, population stable (2016)
Population: Unknown

Comoro Scops-Owl (aka Grand Comoro Scops-Owl, aka Karthala Scops-Owl) (*Otus pauliani*)

Range: Grande Comore (Africa)
Status: Endangered, population decreasing (2020)
Population: 2300

Costa Rican Pygmy-Owl (*Glaucidium costaricanum*)

Range: Costa Rica and Panama
Status: Least Concern, population stable (2016)

Cuban Pygmy-Owl (*Glaucidium siju*)

Range: Cuba
Status: Least Concern, population stable (2016)

Cyprus Scops-Owl (*Otus cyprius*)

Range: Republic of Cyprus
Status: Least Concern, population trend unknown (2019)

Desert Owl (aka Desert Tawny Owl) (*Strix hadorami*)

Range: Arabian Peninsula to Israel and Jordan; eastern Egypt and Sudan
Status: Least Concern, population stable (2016)

Dusky Eagle-Owl (*Ketupa coromandus*)

Range: Pakistan, Himalayas, India, southeast Asia, eastern China
Status: Least Concern, population decreasing (2016)

Eastern Grass-Owl (aka Australasian Grass-Owl) (*Tyto longimembris*)

Range: India, Southeast Asia, Australia
Status: Least Concern, population decreasing (2016)

Eastern Screech-Owl (*Megascops asio*)

Range: North America
Status: Least Concern, population decreasing (2019)

Elf Owl (*Micrathene whitneyi*)

Range: Southwestern United States and Mexico
Status: Least Concern, population decreasing (2020)

Enggano Scops-Owl (*Otus enganensis*)

Range: Enggano Island (Indonesia)
Status: Near Threatened, population stable (2016)
Population: 2500–10,000

Eurasian Scops-Owl (*Otus scops*)

Range: Africa, Middle East, Eurasia
Status: Least Concern, population decreasing (2021)

Fearful Owl (*Asio solomonensis*; aka *Nesasio solomonensis*)

Range: Bougainville (Papua New Guinea), Choiseul and Isabel Islands (Solomon Islands)
Status: Vulnerable, population decreasing (2016)
Population: 2500–10,000

Ferruginous Pygmy-Owl (*Glaucidium brasilianum*)

Range: South America, Middle America, southwestern United States
Status: Least Concern, population decreasing (2022)

Flammulated Owl (*Psiloscops flammeolus*)

Range: North America
Status: Least Concern, population decreasing (2016)

Flores Scops-Owl (*Otus alfredi*)

Range: Flores Island (Indonesia)
Status: Endangered, population decreasing (2016)
Population: 250–2500

Fraser's Eagle-Owl (*Ketupa poensis*)

Range: Central and western Africa
Status: Least Concern, population decreasing (2016)

Fulvous Owl (*Strix fulvescens*)

Range: Southern Mexico to El Salvador
Status: Least Concern, population decreasing (2016)

Golden Masked-Owl (*Tyto aurantia*)

Range: New Britain (Papua New Guinea)
Status: Vulnerable, population decreasing (2018)
Population: 2500–10,000

Grayish Eagle-Owl (*Bubo cinerascens*)

Range: Central Africa
Status: Least Concern, population stable (2016)

Great Gray Owl (*Strix nebulosa*)

Range: Worldwide at northern latitudes
Status: Least Concern, population increasing (2021)

Great Horned Owl (*Bubo virginianus*)

Range: North, Middle, and South America
Status: Least Concern, population stable (2018)

Guadalcanal Owl (*Athene granti*)

Range: Guadalcanal (Solomon Islands)
Status: Near Threatened, population declining
Population: Unknown

Guatemalan Pygmy-Owl (*Glaucidium cobanense*)

Range: Southern Mexico, Guatemala, Honduras
Status: Least Concern, population decreasing (2016)

Halmahera Boobook (*Ninox hypogramma*)

Range: Halmahera, Ternate, and Bacan Islands (Indonesia)
Status: Least Concern, population decreasing (2016)

Hume's Boobook (*Ninox obscura*)

Range: Andaman Islands (India)
Status: Least Concern, population stable (2016)

Indian Scops-Owl (*Otus bakkamoena*)

Range: Pakistan, Himalayan Asia, India, Sri Lanka
Status: Least Concern, population stable (2016)

Itombwe Owl (aka Congo Bay-Owl) (*Tyto prigoginei*, aka *Phodilus prigoginei*)

Range: Central Africa (Democratic Republic of the Congo to Tanzania)
Status: Endangered, population decreasing (2016)
Population: 2500–10,000

Jamaican Owl (*Pseudoscops grammicus*, aka *Asio grammicus*)

Range: Jamaica
Status: Least Concern, population decreasing (2016)

Japanese Scops-Owl (*Otus semitorques*)

Range: Eastern Asia, Japan
Status: Least Concern, population stable (2016)

Javan Owlet (*Glaucidium castanopterum*)

Range: Java and Bali
Status: Least Concern, population stable (2016)

Javan Scops-Owl (*Otus angelinae*)

Range: Java
Status: Vulnerable, population decreasing (2016)
Population: 1500–7000

Least Boobook (*Ninox sumbaensis*)

Range: Sumba Island (Indonesia)
Status: Endangered, population decreasing (2017)
Population: 10,000–20,000

Lesser Masked-Owl (aka Moluccan Masked-Owl) (*Tyto sororcula*)

Range: Buru Island and the Tanimbar Islands
Status: Possibly a subspecies of the Australian Masked-Owl (*Tyto novaehollandiae*) that is restricted to a small range and presumably rare

Lesser Sooty Owl (*Tyto multipuncata*)

Range: Northeastern Australia
Status: Least Concern, population stable (2016)

Long-eared Owl (*Asio otus*)

Range: North America, Eurasia, Middle East, North Africa
Status: Least Concern, population decreasing (2021)

Long-tufted Screech-Owl (*Megascops sanctaecatarinae*)

Range: Southeast Brazil, northeast Argentina, Uruguay
Status: Least Concern, population decreasing (2016)

Luzon Boobook (*Ninox philippensis*)

Range: Philippines
Status: Least Concern, population decreasing (2016)

Luzon Scops-Owl (aka Luzon Highland Scops-Owl) (*Otus longicornis*)

Range: Luzon (Philippines)
Status: Near Threatened, population decreasing (2016)
Population: Unknown

Maghreb Owl (*Strix mauritanica*)

Range: Morocco, northern Algeria, Tunisia
Status: Not yet estimated (according to Birds of the World, Cornell University)

Makira Owl (*Athene roseoaxillaris*)

Range: Makira (Solomon Islands)
Status: Vulnerable, population decreasing (2016)
Population: 2500–10,000

Malaita Owl (*Athene malaitae*)

Range: Malaita (Solomon Islands)
Status: Vulnerable, population decreasing
Population: 2500–10,000 (BirdLife International)

Maned Owl (*Jubula lettii*)

Range: Central Africa
Status: Unknown (categorized as Data Deficient by IUCN)

Manus Boobook (*Ninox meeki*)

Range: Manus and Los Negros Island (Admiralty Islands, Papua New Guinea)
Status: Least Concern, population stable (2016)

Manus Masked-Owl (*Tyto manusi*)

Range: Manus Island (Papua New Guinea)
Status: Possibly a subspecies of the Australian Masked-Owl (*Tyto novaehollandiae*) that is restricted to a small range and presumably rare

Marsh Owl (*Asio capensis*)

Range: Africa
Status: Least Concern, population stable (2021)

Mayotte Scops-Owl (*Otus mayottensis*)

Range: Mayotte Island (Africa)
Status: Least Concern, population stable (2016)

Mentawai Scops-Owl (*Otus mentawi*)

Range: Mentawai Islands (Indonesia)
Status: Near Threatened, population decreasing (2016)
Population: 10,000–20,000

Middle American Screech-Owl (*Megascops guatemalae*)

Range: Middle America
Status: Least Concern, population decreasing (2016)

Minahasa Masked-Owl (*Tyto inexspectata*)

Range: Sulawesi Island (Indonesia)
Status: Vulnerable, population decreasing (2016)
Population: 2500–10,000

Mindanao Boobook (*Ninox spilocephala*)

Range: Southern Philippines
Status: Near threatened, population decreasing (2016)
Population: Unknown

Mindanao Lowland Scops-Owl (aka Everett's Scops-Owl) (*Otus everetti*)

Range: Bohol, Leyte, Samar, Mindanao, Basilan (Philippines)
Status: Least Concern, population decreasing (2016)

Mindanao Scops-Owl (aka Mindanao Highland Scops-Owl) (*Otus mirus*)

Range: Mindanao (Philippines)
Status: Near Threatened, population decreasing (2016)
Population: Unknown

Mindoro Boobook (*Ninox mindorensis*)

Range: Mindoro Island (Philippines)
Status: Vulnerable, population decreasing (2016)
Population: 2500–10,000

Mindoro Scops-Owl (*Otus mindorensis*)

Range: Mindoro (Philippines)
Status: Near Threatened, population decreasing (2016)
Population: 10,000–20,000

Moheli Scops-Owl (*Otus moheliensis*)

Range: Mohéli Island (Africa)
Status: Endangered, population decreasing (2017)
Population: 260

Moluccan Scops-Owl (*Otus magicus*)

Range: Moluccas and Lesser Sunda Islands (Indonesia)
Status: Least Concern, population decreasing (2022)

Mottled Wood-Owl (*Strix ocellata*)

Range: India and Myanmar
Status: Least Concern, population stable (2016)

Mountain Scops-Owl (*Otus spilocephalus*)

Range: Himalayan Asia and Southeast Asia
Status: Least Concern, population stable (2016)

Negros Scops-Owl (aka Visayan Scops-Owl) (*Otus nigrorum*)

Range: Negros and Panay (Philippines)
Status: Vulnerable, population decreasing (2016)
Population: 1000–2500

New Britain Boobook (*Ninox odiosa*)

Range: New Britain and Watom Islands (Papua New Guinea)
Status: Vulnerable, population decreasing (2018)
Population: 10,000–20,000

Nicobar Scops-Owl (*Otus alius*)

Range: Nicobar Islands (India)
Status: Near Threatened, population decreasing (2019)
Population: 2500–10,000

Northern Boobook (*Ninox japonica*)

Range: Eastern and southeast Asia, including Japan
Status: Least Concern, population stable (2016)

Northern Hawk Owl (*Surnia ulula*)

Range: Northern latitudes of North America and Eurasia
Status: Least Concern, population stable (2021)

Northern Pygmy-Owl (*Glaucidium gnoma*)

Range: North America
Status: Least Concern, population slightly increasing (2016)

Northern Saw-whet Owl (*Aegolius acadicus*)

Range: North America
Status: Least Concern, population decreasing (2016)

Omani Owl (*Strix butleri*)

Range: Northern Oman
Status: Least Concern, population stable (2021)

Palau Scops-Owl (*Otus podarginus*)

Range: Palau (Caroline Islands, Micronesia)
Status: Least Concern, population stable (2016)

Pallid Scops-Owl (*Otus brucei*)

Range: Middle East and western Asia
Status: Least Concern, population stable (2021)

Papuan Boobook (aka Jungle Boobook) (*Ninox theomacha*)

Range: New Guinea
Status: Least Concern, population stable (2016)

Papuan Hawk Owl (aka Papuan Owl) (*Uroglaux dimorpha*)

Range: Papua New Guinea
Status: Least Concern, population decreasing (2017)

Pel's Fishing-Owl (*Scotopelia peli*)

Range: Sub-Saharan Africa
Status: Least Concern, population decreasing (2018)

Pemba Scops-Owl (*Otus pembaensis*)

Range: Pemba Island (Tanzania)
Status: Vulnerable, population decreasing (2016)
Population: 3000

Peruvian Pygmy-Owl (aka Pacific Pygmy-Owl) (*Glaucidium peruanum*)

Range: Ecuador and Peru
Status: Least Concern, population stable (2016)

Peruvian Screech-Owl (aka West Peruvian Screech-Owl) (*Megascops roboratus*)

Range: Southwest Ecuador/northwest Peru
Status: Least Concern, population stable (2016)

Pharaoh Eagle-Owl (*Bubo ascalaphus*)

Range: Northwest Africa and Arabian Peninsula
Status: Least Concern, population stable (2016)

Philippine Eagle-Owl (*Ketupa philippensis*)

Range: Philippines
Status: Vulnerable, population decreasing (2016)
Population: 2500–10,000

Philippine Scops-Owl (aka Luzon Lowland Scops-Owl) (*Otus megalotis*)

Range: Luzon, Catanduanes, Marinduque (Philippines)
Status: Least Concern, population decreasing (2016)

Principe Scops-Owl (*Otus bikegila*)

Range: Principe Island (Africa)
Status: Discovered in 2022; status unknown, probably critically endangered
Population: Unknown

Puerto Rican Owl (*Gymnasio nudipes*)

Range: Puerto Rico
Status: Least Concern, population stable (2016)

Rajah Scops-Owl (*Otus brookii*)

Range: Sumatra and Borneo
Status: Least Concern, population decreasing (2018)

Red-chested Owlet (*Glaucidium tephronotum*)

Range: Western and central Africa
Status: Least Concern, population stable (2016)

Reddish Scops-Owl (*Otus rufescens*)

Range: Malay Peninsula, Indonesia
Status: Near Threatened, population decreasing (2016)
Population: Unknown

Rinjani Scops-Owl (*Otus jolandae*)

Range: Lombok Island (Indonesia)
Status: Near Threatened, population stable (2017)
Population: 2500–10,000

Rock Eagle-Owl (*Bubo bengalensis*)

Range: Indian subcontinent
Status: Least Concern, population stable (2016)

Romblon Boobook (*Ninox spilonotus*)

Range: Sibuyan and Tablas Islands (Philippines)
Status: Endangered, population decreasing (2017)
Population: 250–1000

Rote Boobook (*Ninox rotiensis*)

Range: Rote Island (Indonesia)
Status: First described in 1997 from a specimen collected in a mist net in 1990; described as locally common; reclassified as a separate species from Southern Boobook in 2019.

Rufous Fishing-Owl (*Scotopelia ussheri*)

Range: Sierra Leone to Ghana
Status: Vulnerable, population decreasing (2021)
Population: 2500–10,000

Rusty-barred Owl (*Strix hylophila*)

Range: Southeast Brazil, eastern Paraguay, northeast Argentina
Status: Least Concern, population stable (2021)

Ryukyu Scops-Owl (*Otus elegans*)

Range: Ryukyu Islands (Japan), Lanyu Island off southeast Taiwan, Batanes and Babuyan Islands (Philippines)
Status: Near Threatened, population decreasing (2017)
Population: Unknown

Sandy Scops-Owl (*Otus icterorhynchus*)

Range: Equatorial Africa
Status: Least Concern, population stable (2016)

Sangihe Scops-Owl (*Otus collari*)

Range: Sangihe (Indonesia)
Status: Least Concern, population stable (2016)

Sao Tome Scops-Owl (*Otus hartlaubi*)

Range: São Tomé Island (Africa)
Status: Vulnerable, population decreasing (2018)
Population: 250–1000

Seram Boobook (*Ninox squamipila*)

Range: Seram Island (Indonesia)
Status: Least Concern, population decreasing (2016)

Serendib Scops-Owl (*Otus thilohoffmanni*)

Range: Sri Lanka
Status: Endangered, population decreasing (2016)
Population: 150–700

Seychelles Scops-Owl (*Otus insularis*)

Range: Morne Seychellois National Park, Mahé Island, Seychelles
Status: Critically Endangered, population decreasing (2020)
Population: 200–280

Shelley's Eagle-Owl (*Ketupa shelleyi*)

Range: Central and western Africa
Status: Vulnerable, population decreasing (2018)
Population: 1500–7000

Short-eared Owl (*Asio flammeus*)

Range: Widely distributed worldwide
Status: Least Concern, population decreasing (2021)

Simeulue Scops-Owl (*Otus umbra*)

Range: Simeulue Island (Indonesia)
Status: Near Threatened, population stable (2016)
Population: Unknown

Sjöstedt's Barred Owlet (aka Sjöstedt's Owlet) (*Glaucidium sjostedti*)

Range: West-central Africa
Status: Least Concern, population decreasing (2016)

Snowy Owl (*Bubo scandiacus*)

Range: Circumpolar at northern latitudes
Status: Vulnerable, population decreasing (2021)
Population: 14,000–28,000

Socotra Scops-Owl (*Otus socotranus*)

Range: Socotra Island (Republic of Yemen)
Status: Least Concern, population stable (2016)

Speckled Boobook (*Ninox punctulata*)

Range: Sulawesi, Kabaena, Muna, and Butung Islands (Indonesia)
Status: Least Concern, population stable (2016)

Spot-bellied Eagle-Owl (*Ketupa nipalensis*)

Range: Himalayas, Southeast Asia, southern India, Sri Lanka
Status: Least Concern, population decreasing (2018)

Spotted Owl (*Strix occidentalis*)

Range: Western North America, including Mexico
Status: Near Threatened, population decreasing (2020)
Population: 15,000

Sri Lanka Bay-Owl (*Phodilus assimilis*)

Range: Sri Lanka and southwestern India
Status: Least Concern, population stable (2016)

Subtropical Pygmy-Owl (*Glaucidium parkeri*)

Range: Ecuador, Peru, northern Bolivia
Status: Least Concern, population stable (2016)

Sula Scops-Owl (*Otus sulaensis*)

Range: Sula Island (Philippines)
Status: Near Threatened, population stable (2016)
Population: Unknown

Sulawesi Masked-Owl (*Tyto rosenbergii*)

Range: Indonesia
Status: Least Concern, population stable (2016)

Sulu Boobook (*Ninox reyi*)

Range: Sulu Archipelago (Philippines)
Status: Vulnerable, population decreasing (2016)
Population: 1000–2500

Sumba Boobook (*Ninox rudolfi*)

Range: Sumba and Lesser Sunda Islands (Indonesia)
Status: Near Threatened, population decreasing (2016)
Population: 6000–15,000

Sunda Owlet (*Taenioptynx sylvaticus*)

Range: Sumatra and Borneo
Status: Least Concern, population decreasing (2021)

Taliabu Masked-Owl (*Tyto nigrobrunnea*)

Range: Taliabu Island (Indonesia)
Status: Vulnerable, population trend unknown (2016)
Population: 250–1000

Tamaulipas Pygmy-Owl (*Glaucidium sanchezi*)

Range: Northeastern Mexico
Status: Near Threatened, population decreasing (2019)
Population: 20,000–50,000

Tasmanian Boobook (*Ninox leucopsis*)

Range: Tasmania
Status: Least Concern, population stable (2016)

Tawny Fish-Owl (*Ketupa flavipes*)

Range: Himalayas to Southeast Asia
Status: Least Concern, population decreasing (2018)

Tawny-browed Owl (*Pulsatrix koeniswaldiana*)

Range: Brazil, Paraguay, Argentina
Status: Least Concern, population decreasing (2018)

Timor Boobook (*Ninox fusca*)

Range: Timor, Roma, Leti, and Semau Islands (Indonesia)
Status: Unknown; reclassified as a separate species from Southern Boobook in 2019

Togian Boobook (*Ninox burhani*)

Range: Togian Islands (Indonesia)
Status: Near Threatened, population decreasing (2016)
Population: 1500–7000

Vermiculated Fishing-Owl (*Scotopelia bouvieri*)

Range: West-central Africa
Status: Least Concern, population decreasing (2018)

Vermiculated Screech-Owl (*Megascops vermiculatus*)

Range: Middle America
Status: Least Concern, population decreasing (2016)

Verreaux's Eagle-Owl (*Ketupa lacteus*)

Range: Africa
Status: Least Concern, population stable (2016)

Wallace's Scops-Owl (*Otus silvicola*)

Range: Sumbawa and Flores (Indonesia)
Status: Least Concern, population stable (2016)

West Solomons Owl (*Athene jacquinoti*)

Range: Solomon Islands
Status: Least Concern, population probably decreasing

Western Screech-Owl (*Megascops kennicottii*)

Range: North America
Status: Least Concern, population decreasing (2016)

Wetar Scops-Owl (*Otus tempestatis*)

Range: Wetar Island (Indonesia)
Status: Least Concern, population stable (2016)

Whiskered Screech-Owl (*Megascops trichopsis*)

Range: Arizona and Middle America
Status: Least Concern, population increasing (2016)

White-fronted Scops-Owl (*Otus sagittatus*)

Range: Malay Peninsula north to Myanmar/Thailand border region
Status: Vulnerable, population decreasing (2016)
Population: 2500–10,000

Acknowledgments

This book would not have been possible without the support of many outstanding photographers from around the world who shared their work for the asking. I am especially indebted to Nick Athanas, Tania Simpson, Ron Wolf, Claude Allaert, Isaac Clarey, Hilary Bralove, Ellen Jo Roberts, Jeff Maw, Matt Grube, James Eaton, Alejandro Cartagena, Bruno Conjeaud, Jay Koolpix, Chuck Gates, Marina Scarr, and Vincenzo Penteriani. As the photo credits show, this book includes dozens of images sourced through Creative Commons and I sincerely thank all these photographers and artists for making their captivating work available for public and commercial work.

I am also profoundly grateful for the highly informative input from owl rehabilitators from numerous nations. These people are amazingly dedicated and highly inspiring. Thanks also to Steve Tolleson, my partner in physically demanding boots-busting forays deep into remote desert environs where wild birds roam, for his frequent and effective badgering and berating to keep me on top of deadlines; to Liz Kiren for her superb proofreading and nonstop support; to the Timber Press family for its commitment to excellence, and especially to my intrepid acquisitions editor from Timber Press, Will McKay, for his highly efficacious guidance of this project from inception to completion.

Photo and Illustration Credits

All photographs by John Shewey with the exception of the following:

Bruno Conjeaud, 207 (bottom, left)
Cadop via Pixibay, 50
Claude Allaert, 135
CTolman via Pixibay, 113 (left)
ed_mcaskill via Pixibay, 108 (top, left)
Ellen Jo Roberts, 33 (left)
Erik Karits via Pixibay, 108 (top, right; bottom)
Hilary Bralove, 102 (middle, right)
Internet Archive/*The smugglers; picturesque chapters in the story of an ancient craft.* By Charles George Harper, 1909. Page 36
James Eaton, 214 (bottom, right)
Jay Koolpix, 27 (right)
Kate Dolamore, 79
Kellee Abell, 173
Kyle Brooks/Forest Service, 130
Marina Scarr, 41
Mark K. Daly, 47 (left)
Mike Budd/USFWS Public Domain, 147
National Park Service/Santa Monica Mountains National Recreation Area, 172
Nick Athanas, 57 (bottom), 195, 196, 199 (bottom, left), 200 (bottom, right), 203 (top, left; bottom), 204 (bottom, left),205 (top, right), 206 (bottom, right), 207 (bottom, right), 208 (top, left), 211 (top, left; bottom), 212 (bottom, right), 213 (top, left), 214 (top, left), 215 (top, right; bottom, left), 216 (bottom), 217

Ron Wolf, 14

Smithsonian Libraries and Archives/Biòdiversity Heritage Library/Vccelliera, overo, Discorso della natvra, e proprieta di diversi vccelli. By Giovanni Pietro Olina 1622, Anotnio Tempesta (1555-1630), Francesco Vilamena (1566-1624). Page 44

Suan Hsi Yong, 160

Tania Simpson via Pixibay, 18, 111 (bottom, left), 125 (top, left), 162

Tereza Houdová, 12

The Metropolitan Museum of Art/"Terracotta lekythos (oil flask)" Attributed to the Brygos Painter. Rogers Fund, 1909. Accession Number: 09.221.43. Page 21 (left)

Vincenzo Pentariani, 38

Flickr

Alejandro Cartagena, 199 (bottom, right)

Isaac Clarey, 200 (bottom, left), 201 (top, left)

James A. Lawson, 77, 126

Jeff Maw, 93 (top, left), 97 (bottom), 209 (bottom, left)

Jim Griffin, 68

Matt Grube, 201 (bottom)

CC BY 2.0

Andrey Gulivanov, 25, 100 (top, right), 106, 111 (bottom, right), 205 (top, left), 215 (top, left), 216 (top, left

Andy Morffew, 88 (top), 207 (top, right)

Andy Reago & Chrissy McClarren, 59, 60, 80 (top, right), 83 (top, left), 119 (left), 144

Becky Matsubara, 96, 105 (top, right)

Bill Morrow, 17

Caroline Legg, 80 (top, left), 105 (bottom), 164

Channel City Camera Club/Steve Colwell, 125 (bottom)

Charles Gates, 125 (top, right)

Chuck Gates, 111 (top)

Daniel Voyager, 54

David Lochlin, 210 (bottom, right)

Derek Keats, 49, 199 (top), 210 (bottom, left)

Eduardo Schmeda, 200 (top, right)

Elusive Muse, 29

Fyn Kynd, 100 (top, left)

Greg Schechter, 85

Hari K. Patibanda, 206 (top, right)

Henry Weinberg, 69

Henry, 42

Jason Thompson, 209 (top, right)

Jo Garbutt, 71

Koji Ishii, 62

Kyle Sullivan, Bureau of Land Management, 116 (bottom)

Larry Lamsa, 57 (bottom)

Lorie Shaull, 99

Mark Braggins, 80 (bottom)

Martinus Scriblerus, 30

Mount Rainier National Park, 117

National Museum of Denmark, 65
nature80020_Jim Kennedy, 70
Nigel, 118
Richard, 192
Robert Claypool. 47 (right)
Russell Chilton, 208 (bottom, left)
sam may, 113 (bottom, right)
shankar s, 205 (bottom, right)
Silver Leapers, 1231 (top)
Stephane Tardif, 204 (bottom, right)
Terry Wing, 19
The Cut, 28
Tom Brandt, 33 (right)
USFWS Mountain-Prairie, 102 (bottom, right; bottom, left)
Virginia State Parks, 159

CC BY ND 2.0

Ashley Wahlberg (Tubbs), 93 (bottom)
Chris Foster, 23
Heckrodt Wetland Reserve, 83 (bottom, right)
Melanie C. Underwood, 90 (top, right)
Tambako the Jaguar, 4, 24
Tom Spine, 90 (bottom, left), 193
vil.sandi, 72

CC BY SA 2.0

Allen Gathman, 168 (right)
Bernard DuPont, 95
Chris McCafferty—Oregon State University, 115 (top)
cuatrok77, 83 (bottom, left), 137
DaPuglet (Tina), 52 (right)
Deborah Freeman, 56
Dominic Sherony, 90 (bottom, right), 93 (top, right), 136
Gerry Zambonini, 216 (top, right)
Graham Winterflood, 211 (top, right)
Joris Komen, 198
Kameron Perensovich, 97 (top), 113 (top, right)
Kimon Berlin, 207 (top, left)
Matt Lavin, 168 (left)
Mike W, 45
russellstreet, 8

Public Domain

Alaska Region US Fish and Wildlife Service, 184
Dani Ortiz, Bureau of Land Management, 52 (left)
Debbie Koenigs/USFWS, 185 (bottom)
Hailey Frost—Lincoln National Forest, 116 (top)
Jake Bonello—USFWS, 102 (top, right)
Ken Sturm/USFWS, 179
Lisa Hupp/USFWS Alaska Region, 34 (bottom)
NPS/N. Lewis/Shenandoah National Park, 180
Peter Pearsall/USFWS Alaska Region Flickr, 51
Seth Topham—Bureau of Land Management, 105 (top, left)
Shenandoah National Park, 27 (left)
USFWS Pacific Southwest, 115 (bottom)

Unsplash

Pete Nuij, 140
Richard Lee, 119 (right)

CC by Unsplash

Doug Kelley, 131 (right)
Gary Bendig, 131 (left)
Zedenek Machacek, 121 (bottom)

Wikimedia

CC BY 2.0

Hans Norelius, 210 (top, left)
Jason Thompson, 212 (bottom, left), 214 (top, right)
Ken and Nyetta, 202 (bottom, right)
Kevin Cole, 22
Michael Gillam, 202 (bottom, left)
Pai-Shih Lee/白士 李, 204 (top, right)

CC BY 2.5

Museum of Fine Arts of Lyon. Found in the area of Lyon. Photographer: Marie-Lan Nguyen. Page 21 (right)

CC BY SA 2.0

Imran Shah, 86 (top), 204 (top, left)
Joachim Huber, 209 (top, left)
Kameron Perensovich, 34 (top)
travelwayoflife, 88 (bottom)

CC BY SA 3.0

JJ Harrison, 212 (top, left)
Mdf, 83 (top, right)

CC BY SA 4.0

Abhijit Sarma, 200 (top, left)
Allan Barredo, 208 (top, right)
Chuck Martin, 90 (top, left)
Davidvraju, 209 (bottom, right)
Dfpindia, 201 (top, right)
Dominic Sherony, 208 (bottom, right)
Frank Schulenburg, 87
Fyn Kynd, 127, 128
Infrogmation of New Orleans, 181
JJ Harrison, 212 (top, right), 213 (bottom, left)
John Mwacharo, 187
Josikk, 100 (bottom)
Mikael Bauer, 171
Mousam ray, 206 (top, left)
Nortondefeis, 202 (top), 213 (bottom, right), 215 (bottom, left)
Punitha Satharasinnghe, 203 (top, right)
Raafi Nur Ali, 214 (bottom, left)
Rawpixel Illustration from Svenska Fåglar, 53
Santanu Majumdar, 58
Saswat Mishra, 205 (bottom, left)
thibaudaronson, 206 (bottom, left)
Timothy A. Gonsalves, 213 (top, right)
Wosella, 86 (bottom)
Zoran Gavrilovic, 35

Public Domain

National Library New Zealand, 174
National Palace Museum, 63
Unknown, 191

ZooKeys

CC BY 4.0

Melo M, Freitas B, Verbelen P, da Costa SR, Pereira H, Fuchs J, Sangster G, Correia MN, de Lima RF, Crottini A (2022) A new species of scops-owl (Aves, Strigiformes, Strigidae, Otus) from Príncipe Island (Gulf of Guinea, Africa) and novel insights into the systematic affinities within Otus. ZooKeys 126:1-54. https://doi.org/10.3897/zookeys.1126.87635; photo by Philippe Verbelen. Page 66

Bird Maps from Cornell

From *The Birds of The World*, birdsoftheworld.org, published by the Cornell Lab of Ornithology, 2019.

Resources

Useful Books about Owls

Barn Owls: Evolution and Ecology by Alexandre Roulin. Cambridge: Cambridge University Press, 2020.

Bird Gods by Charles de Kay and George Wharton Edwards. New York: A.S. Barnes & Company, 1898.

The Eagle Owl by Vincenzo Penteriani and Maria del Mar Delgado. Staffordshire, UK: T & AD Poyser, 2019.

The Great Horned Owl: An In-depth Study by Scott Rashid. Atglen, PA: Schiffer Publishing, 2015.

Owls of the Eastern Ice: A Quest to Find and Save the World's Largest Owl by Jonathan C. Slaght. New York: Farrar, Straus and Giroux, 2020.

The Screech Owl Companion: Everything You Need to Know about These Beneficial Raptors by Jim Wright and Scott Weston. Portland: Timber Press, 2023.

The World of Burrowing Owls: A Photographic Essay Exploring Their Behaviors & Beauty by Rob Palmer. Amherst Media, 2019.

Owl Research and Advocacy Organizations

All About Owls

Petaluma, California
(415) 454-4587
www.allaboutowls.org

American Bird Conservancy

www.abcbirds.org

Burrowing Owl Conservation Network

(Urban Bird Foundation)
American Canyon, California
(925) 240-3399
www.burrowingowlconservation.org

Burrowing Owl Conservation Society of BC

British Columbia, Canada
www.burrowingowlbc.org

Burrowing Owl Preservation Society

Woodland, California
burrowingowls.org

Cornell Lab of Ornithology

Ithaca, New York
(607) 254-2473
www.birds.cornell.edu

Hungry Owl Project

San Rafael, California
www.hungryowls.org

International Owl Center

Houston, Minnesota
(507) 896-6957
www.internationalowlcenter.org

Manitoba Burrowing Owl Recovery Program

Manitoba, Canada
(204) 807-4668
www.mborp.ca

Owl Research Institute

Charlo, Montana
(406) 644-3412
www.owlresearchinstitute.org

Partners in Flight

www.partnersinflight.org

Project SNOWstorm

www.projectsnowstorm.org

Saskatchewan Burrowing Owl Interpretive Center

Moose Jaw, SK, Canada
(306) 692-8710
www.skburrowingowl.ca

The Owl Foundation

Ontario, Canada
www.theowlfoundation.ca

The Peregrine Fund

(The World Center for Birds of Prey)
Boise, Idaho
(208) 362-3716
www.peregrinefund.org

The Raptor Resource Project

Decorah, Iowa
(276) 325-2662
www.raptorresource.org

Index

B

C

D

O

T

About the Author

Lifelong birding enthusiast **John Shewey** is a professional writer, editor, and photographer whose work has appeared in dozens of magazines, covering topics ranging from fly-fishing and travel to natural history and wildlife to cuisine and canines. Among his nearly two dozen books are popular favorites such as *The Hummingbird Handbook*, *Birds of the Pacific Northwest*, *Oregon Beaches: A Traveler's Companion*, *Spey Flies: Their History and Construction*, and *Classic Steelhead Flies*. John rejoices in the outdoors, is addicted to fly angling, and enjoys studying wildlife; he is a desert rat who has left boot prints in Oregon's most remote places, and as a dog aficionado he has owned people-loving Weimaraners for 30 years; he revels in tiny towns, excellent whisky, good wine, decent ales, remote campsites, and back-country roads.